The two data processing sides of our upper brain are separated by a deep cleft because they are doing opposite things that cannot be done in the same space. The left hemisphere, "Splitter", is doing top-down analysis of the important details. In contrast, the right side, "Lumper" is doing bottom-up analysis of the big picture.

Since there can be only one Executive in a bilateral brain, we were each born with our executive located either in the left or the right hemisphere. This makes us inherently either a right brain-oriented person (RP) or left brain-oriented person (LP). This topic is called Hemisity.

Recently, several methods have been developed to determine a person's hemisity. It was found that LPs and RPs differ upon how they look at the world. LPs, like the left hemisphere are important details splitters, and RPs, like the right hemisphere are big picture lumpers.

In heterosexual courtship and marriage, there are four possible hemisity combinations: LM-RF, RM-LF, RM-RF and LM-LF (where M and F are male and female). Referring to hemisity, it was found that overall, "opposites attract" in mate selection.

Anciently, this resulted in two complementary "true-breeding" species with their many unrecognized historical conflicts. These two species are <u>Homo sapiens matripolaris</u> (LM-RF couples) and <u>Homo sapiens patripolaris</u> (RM-LF couples).

In contrast, the same-same, RM-RF, and LM-LF marriages are cross-breed matches between the matri and patripolar mates. Unavoidably, their children are interspecies hemisity hybrids.

In hemisity hybrids, there is the contribution of one set of hemisity genes from the mother from one species, with that of a different set from the father of the other species. As a result in hybrids, there are developmental factor deficits or excesses during the second trimester of pregnancy. *These result in erroneous crossings during the formation of the midline of the central nervous system.* This causes the reversal of sexual identity, or partner preference in certain specific hybrids. In others, it leads to irreversible dyslexia or schizophrenia.

It is my belief that the widespread communication of the discovery of two human species and their hybrids will ultimately reduce human suffering.

"Dr. Morton has provided us a monumental paradigm-shift regarding who we are as humans. Its effects will be enormous." DENNIS G. McLAUGHLIN, Ph.D., Clinical Psychologist, Care Hawaii, Honolulu, HI.

"Dr. Morton's book shines new thinking on the human brain. Particularly intriguing is his assessment of the hemisity of Homo sapiens and of how this explains the context of culturally significant, controversial behavioral abnormalities. His assertions are strongly documented and rigorously argued. Anyone interested in gaining a unique perspective of the relationship of brain to behavior would be well served by reading this important work." KATHRYN KO, M.D., MFA, Chief, MetroNeurosurgery, New York, NY.

"Even if only half of Dr. Morton's observations turn out to be empirically supported, he will still have made an unparalleled contribution to the understanding of human behavior at many levels." MICHAEL P. KELLEY, Ph.D., Clinical Psychologist, Washington, D.C., USA.

"A great deal of misinformation exists about topics related to human sexuality-to say nothing of ongoing controversies that embroil families, churches, and belief systems—so that anything or anyone who can shine the light of increased knowledge on the subject is most welcome!"
ARLENE R. TAYLOR, Ph.D., Brain Function Specialist
Founder/President of Realizations, Inc.

"In this fascinating book Professor Bruce Morton presents his revolutionary idea about previously unknown hereditary mechanisms that define human behavior. He targets the most elusive element of human consciousness, the Self. Without doubt, the book opens a new era in psychology, and would be most useful for clinical psychologists and cognitive neuroscientists; yet it is written simply enough to be appreciated by anybody interested in self-knowledge." EUGENE NALIVAIKO, Ph.D. Neurocardiologist, University of Newcastle, Callaghan, Australia

"Dr. Morton should be nominated for the Nobel Peace Prize for this brilliant and thought-provoking work. Not since Darwin has such a world-changing wealth of new ideas come to challenge our knowledge of who we are." E. A. HANKINS III, MD, Curator of Vertebrate Zoology, Founder of World Museum of Natural History, Riverside, CA.

"Scientists, philosophers, and politicians have long struggled to deal with the conundrum of paradoxical behavior. Bruce Morton has provided a paradigm in his latest work to provide one means of untying this Gordian knot. You may find some of the premises unsettling, but no advance in thinking has happened without things being shaken up." ROBERT C. MARVIT, M.D., Neuropsychiatrist, President Hawaii Medical Association

"Professor Bruce Eldine Morton's *Two Human Species Exist* is an amazing and paradigm-shifting work. After reading it, the world will look different to you. Isn't it at least possible that life emerged more than once on this planet, and that evolution progressed on at least two parallel tracks Isn't it at least possible that the differences between people and peoples can be explained by there being two different species of humans? Isn't it at least possible that interbreeding between the two species results in hybrids, which may have advantages and also flaws not possessed by the purebreds of the two species? Professor Morton's ideas will challenge you on every page and make you see humans and, indeed, the world in new ways." HAROLD LENHART, M.D., Assistant Clinical Professor of Psychiatry, Michigan State University College of Human Medicine.

Also by Bruce Morton:

NEUROREALITY: A Scientific Religion to Restore Meaning

Megalith Books, 2011 (amazon.com)

PSYCHEDELIC VISIONS FROM THE TEACHER:

A Neuroscientist's Initiation to Reality and Spirituality

Megalith Books, 2013 (amazon.com)

BEYOND MEN ARE FROM MARS

Megalith Books, 2014 (amazon.com)

MODULAR CONSCIOUSNESS

Megalith Books, 2015 (amazon.com)

http://www2.hawaii.edu/~bemorton

lists Dr. Morton's neuroscience publications.

TWO HUMAN SPECIES EXIST

Their Hybrids Are Dyslexics, Homosexuals, Pedophiles, and Schizophrenics

Bruce Eldine Morton

Megalith Books

Doral, Florida

COPYRIGHT

TABLE OF CONTENTS

DEDICATION:

To Roger Deduvir, M.D. Emeritus Professor, Einstein College of Medicine for his insight and encouragement.

ACKNOWLEDGEMENTS:

This book represents an expansion of four chapters in the book Neuroreality. These ideas are the logical extension of my unfunded research findings on human hemisity, the forbidden topic.

PREFACE: A Hybrid's Quest For Normalcy

I was born a developmental dyslexic. But, like the majority of us hybrids, I didn't know that I was cross-wired. The reason I didn't know this until I was in my late 50s was because I was also born very intelligent. People with strong parental support that work hard can almost compensate for the impaired memory that comes with dyslexia. They can kid themselves that they are normal. Most people believe that we are created equal. Actually, this is very far from the truth. We may have politically correct equal rights under the law, but we are not equal, as any glance at the Guinness Book of Records will affirm.

The hardest period for the millions of people with developmental dyslexia, or the other psychosocial arrests, such as homosexuality, occurs during their early years in school. My spelling was atrocious, my handwriting was horrible, learning required the endless rote repetition of memorization, and arithmetic came very hard. I was near the bottom of the class and beginning to show ego-saving behavioral problems. What I didn't know was that normal "wild type" people (genetically speaking) naturally had photographic memories. To see something once was all they needed to remember it forever; no memorization required. Since I believed that we were all equal, I ascribed their easy success in school to their harder work and greater diligence. I punished myself as lazy and undisciplined. I vowed to work harder.

Somehow, I did learn to read and began to bring a dozen books home from the library every two weeks. The library was a wonderful place! This was the time before television was available. Books became my window to the world. I became as addicted to reading books then, as a kid today is to games on his playstation.

Since my high school overall grade point average (GPA) was a low 2.6 out of 4.0, I was fortunate to be able to enter a local college. Then the crisis came. I had no idea whether I could do college level work. Very concerned, I went to the Student Counseling Office, and they gave me tests, including an IQ test. Even with my unrecognized dyslexia, my IQ was in the top 2%. They said that I could certainly be successful in college if I worked hard.

With that weight off my back, I enrolled in a chemistry major curriculum. Although most of my friends were premedical, a bright girlfriend whispered in my ear: "medical doctors heal their thousands, but medical scientists improve the lives of millions." I wanted to become a scientist!

For my first year General Chemistry Course, I received a grade of a B, despite being unable to answer any questions on the exam that required math. Amazingly, during my second year while taking Quantitative Analysis, something unexpected happened. I suddenly understood and could do the math needed to analyze and solve each problem. Later, I passed two required Calculus courses with no difficulty. In retrospect, it is clear late adolescence brings several unrecognized brain development steps that at the time were invaluable to me (but that go wrong in schizophrenics). I graduated from college with a Bachelor of Arts and a GPA of a modest 3.2. I still felt stupid because of all the effort it took me, while others with higher grades just seemed to play.

I began to blossom and obtained a M.S. and Ph.D. degrees from the University of Wisconsin, then ranked #5 in the world in my field. This time, my overall GPA was 3.9. This was followed by postdoctoral work at M.I.T. and Harvard Medical School that resulted in a number of publications. I still felt stupid, but now I was beginning to understand how to level my playing field. Einstein was also dyslexic, with delayed speech and problems with his grammar school math. He too was cross-wired.

To work up through the faculty ranks at the University of Hawaii required enormous effort on my part to compensate for my still unrecognized dyslexia. It was so stressful that I developed seasonal depression that appeared every fall with the beginning of the new academic year. To abbreviate the story told in the preface and last chapter of my book "Neuroreality", I ultimately used hallucinogenic substances to undergo repeated "near death experiences". These many ego deaths had a profound transformative effect upon me. It opened my eyes and led to my discovery of many things, including Familial Polarity, the crucially important subject of this book. Oh yes, when I retired from the University of Hawaii, it was at the rank of Professor Emeritus. The stupid hybrid had made it.

INTRODUCTION

INTRODUCTION: RE-CREATING THE UNIVERSE

The time has come when we have more than enough information about the universe, earth, life, brain, mind, and spirit to enable a more accurate reformulation of where and who we are. Our traditional universe model is inaccurate and more than 4,000 years obsolete. The key element in such a recreation of the universe was the complete exclusion of all supernatural processes. Such a reformulation, called *Neuroreality: A Scientific Religion To Restore Meaning,* has already been published (93). It includes the following 11 key elements:

1. The universe is infinite, eternal, and filled with endless webs of galaxies. All big bangs are local.

2. As **Figure 1** illustrates, galaxies are shaped like two reverse stacked pumpkins. However, we only see the spiraling inward stars and nebulae of the central plane, imperceptively moving toward the black hole at the core. The rest of the invisible stacked double tori consists of dark energy and subatomic dark matter ejected from the black hole singularity by continuous polar big bang expansions.

3. As the erupting big bang dark energy transforms into subatomic dark matter, it is bent around by the enorMous gravity of the galaxy. From above an dbelow, these enter the planar central spiral and become visible as luminous hydrogen nebulae become the first light matter.

4. Within the planar arms of galaxies like ours is a

Figure 1:

Galactic Engine Model: Origin of Life

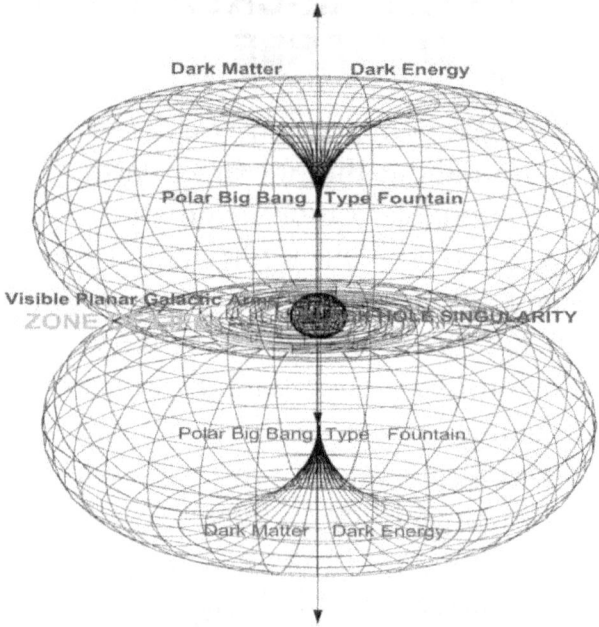

Dark Matter   Dark Energy

Polar Big Bang Type Fountain

Visible Planar Galactic Arm   ZONE   HOLE SINGULARITY

Polar Big Bang Type Fountain

Dark Matter   Dark Energy

m

matter/energy balanced circular "halo of life" through which all stars must pass on their way toward the galactic center where no stars are formed.

5. As the stars and their planets pass into such fertile zones, cellular life spontaneously originates through chemical evolution and evolves to such complexity as to become sentient, only to be destroyed eons later upon leaving the protection of the halo. Upstream is the direction we should be looking for life-supporting planets still within the halo.

6. But, because of the huge distances involved, we may never be aware of life forms occurring in the other arms of our galaxy, nor probably those originating even in our own galactic arm within the zone of life.

7. Here on earth, however, it is becoming clear that life has originated more than once. Wearing the blinders of our childhood model of the universe, even Darwin didn't conceive of this possibility. However, his pioneering evolutionary model of "survival of the fittest" was required to set the stage for perception of this step.

8. The multiple origins of life on earth are supported by considerable evidence, including that for two existing but yet unrecognized trees of life.

9. The animals of each of these two trees use opposite reproductive strategies, one harem forming and the other territorial. Contrasting examples of this dichotomy are seen in wolves vs. foxes, horses vs. donkeys, rats vs. mice, etc.

10. As a result, humanity is composed of two genetically different patripolar and matripolar pre-racial species, whose hybrids show cross-wiring defects.

11. The formation of these hybrids provides the first rational explanation accounting for the production of dyslexia, bisexuality, homosexuality, pedophilia, gothic orientations, and schizophrenia.

This book enables you to obtain hands-on evidence for yourself that supports items 7 through 11 above.

Are the Ideas in This Book True?

First, it must be said that the new concepts and contexts presented in this book are arbitrarily treated here as facts. Many become so obviously true that you will soon begin to find convincing support for them almost everywhere. However, because most of these ideas are new, the time, rigor, and replication required for formal scientific proof are impossible to provide at present, and are perhaps is a century away. Certain details may be found to be in error, but the overall structure of this new paradigm will survive.

Second, because these concepts and contexts are new, no scientist or other authority is presently familiar with them. Because they challenge tradition, their first reaction may be negative. Many among them will think, "I know at least something about most everything. If I haven't heard of this, then it doesn't exist." Or, "It's just plain wrong." That will be because these ideas are not yet part of our cultural consensual reality. But, once they are named and defined, as in this book, they will become obviously true, and later science will be declare them so.

CHAPTER 1. Evidence Part 1: DISCOVER YOUR HEMISITY

Discovering Hemisity Will Transform Your Relationships Forever

Although sex or handedness has been used to segregate people into subgroups, this book confirms the existence of another binary difference that also sorts people into two different groups, independent of handedness and sex. These are the right brain-oriented or left brain-oriented behavioral subgroups of Hemisity. The right brain-left brain idea is not new. In fact, awareness of the laterality of brain function appears to be at least as old as written history. For example, Diocles of Carystus in the fourth century BC wrote: "There are two brains in the head, one which gives understanding, and another which provides sense-perception. That is to say, the one which is lying on the right side is the one that perceives: with the left one, however we understand" (76).

In recent times, right and left-brain differences were popularized as a result of the Nobel Prize winning split-brain studies of Roger Sperry (114-116). These procedures were performed because it had been found that surgically disconnecting cerebral hemispheres could block grand mal seizures of epileptics. This was done by an operation called a callosotomy in which the coaxial cable between both sides of the brain, called the corpus callosum, was cut (7-12, 43-47, 98, 114-116). This worked by preventing the spread of seizure-producing excitation from one side of the brain to the other. It is still used in some cases today.

In these recovered and normally acting split-brain patients, means were found to communicate separately with each of the hemispheres. Strikingly, the hemispheres said different things and seemed to have different thinking and behavioral styles (43-47, 115-116). The question became, which hemisphere manifests in normal persons? Then, people began to notice that some non-epileptic persons appeared predominately to show many of the thinking and behavioral styles of the right hemisphere, while others seemed more left hemisphere like. This concept was called hemisphericity (43-47), and interest in it resulted in thousands of research papers on the topic (24, 59, 61, 96).

The Bottom-Up and Top-Down Processing of the Two Cerebral Hemispheres Requires Physical Separation

The functionally opposite orientation of the two powerful data-abstracting units within the cerebrum (35, 62, 73, 105) is shown in **Figure 2**, using one of the figures describing the Dual Quadbrain Model found in "Neuroreality" (93). In brief, the left cerebral hemisphere (LH) sees differences between things and uses top-down, deductive reasoning from the general to the particular to dissect the next lower-universe level. It is an analytical, intelligent "Splitter" with an orientation toward *content*. It focuses upon the trees, not forests, facial elements, rather than faces (55, 117, 146).

In contrast, the *context* orientation of the right hemisphere (RH) assists it in the detection of global similarities in an intuitive and at times metaphorical way (31, 110, 143-145). It uses inductive reasoning to go from the particular (individual) instances to the general (group) commonality in bottom-up thinking, to synthesize the next higher universe-level, and thus is a "Lumper". It sees the forest over the trees and is good at recognizing faces. Thus, the orientation of the RH is for visual, concrete images where "a *picture* is worth a thousand *words*". This contrasts with orientation of the LH toward abstractions, where "a *word* is worth a thousand *pictures*", an often overlooked fact.

The two hemispheres must remain separate because the two opposite processes performed by the hemispheres are incompatible. However, the two exchange limited amounts of information via corpus callosum, and the deeper sub cortical anterior and posterior commissures. Because of their differences, each cerebrum performs mutually exclusive, survival-maximizing data processing operations.

In the right brain, incoming data is *inductively* compared (with the assistance of the striatal matching system) with earlier-similar memory data to see whether the two data sets might be *similar and related*. It is of great survival value to know rapidly if both sets of data are related. If so, earlier-similar outcome memories can next be scanned in terms of past survival harm or benefit. Then avoidance or approach behavior is initiated, seeking to increase the survival benefit of the present situation.

In exclusive contrast, in the left-brain the incoming data is *deductively* compared with "earlier-similar"

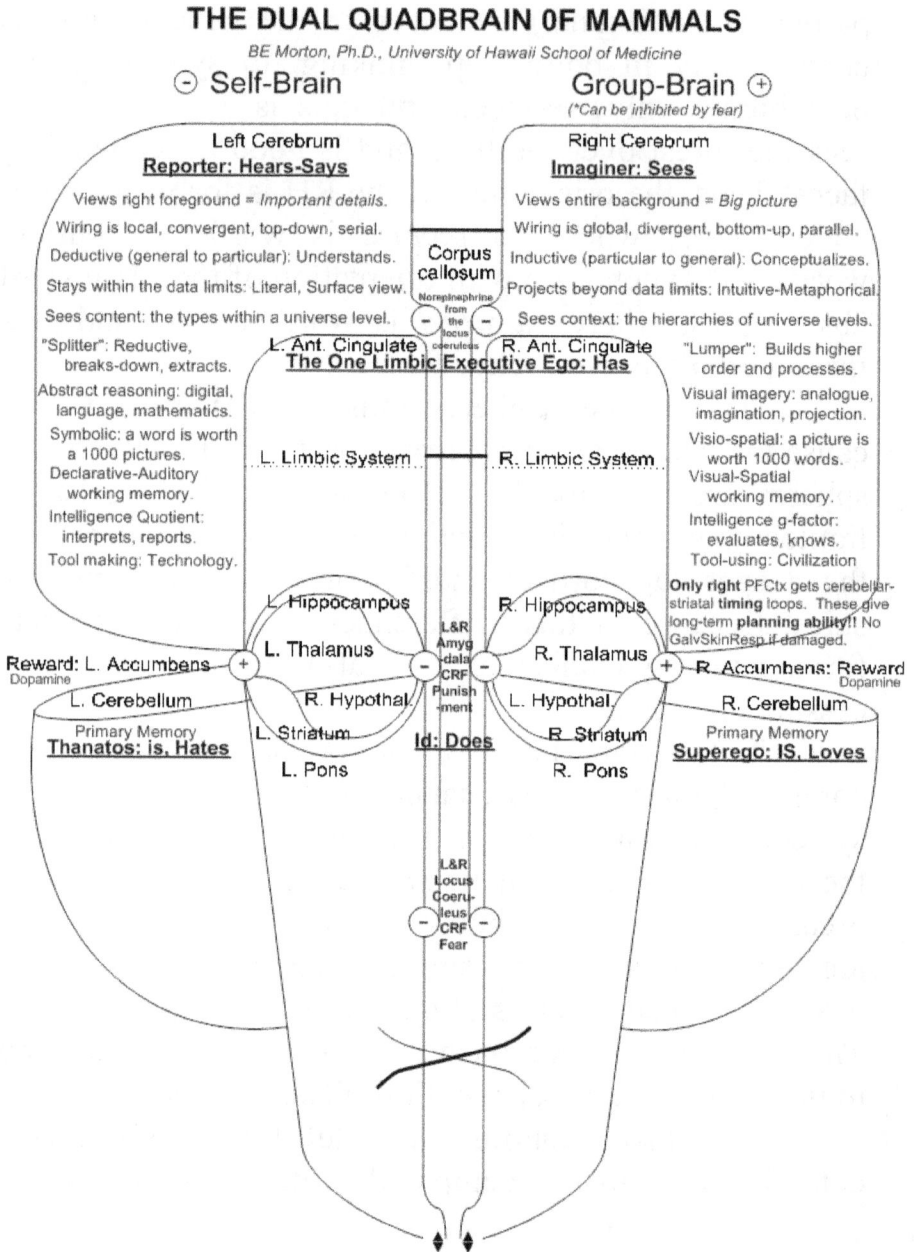

Figure 2:

THE DUAL QUADBRAIN OF MAMMALS

BE Morton, Ph.D., University of Hawaii School of Medicine

⊖ Self-Brain   Group-Brain ⊕

(*Can be inhibited by fear)

Left Cerebrum **Reporter: Hears-Says**		Right Cerebrum **Imaginer: Sees**
Views right foreground = *Important details.*		Views entire background = *Big picture*
Wiring is local, convergent, top-down, serial.	Corpus callosum	Wiring is global, divergent, bottom-up, parallel.
Deductive (general to particular): Understands.		Inductive (particular to general): Conceptualizes.
Stays within the data limits: Literal, Surface view.	Norepinephrine from the locus coeruleus	Projects beyond data limits: Intuitive-Metaphorical.
Sees content: the types within a universe level.		Sees context: the hierarchies of universe levels.

L. Ant. Cingulate ⊖ ⊖ R. Ant. Cingulate
The One Limbic Executive Ego: Has

"Splitter": Reductive, breaks-down, extracts.		"Lumper": Builds higher order and processes.
Abstract reasoning: digital, language, mathematics.		Visual imagery: analogue, imagination, projection.
Symbolic: a word is worth a 1000 pictures. Declarative-Auditory working memory.	L. Limbic System R. Limbic System	Visio-spatial: a picture is worth 1000 words. Visual-Spatial working memory.
Intelligence Quotient: interprets, reports.		Intelligence g-factor: evaluates, knows.
Tool making: Technology.		Tool-using: Civilization

Only right PFCtx gets cerebellar-striatal timing loops. These give long-term **planning ability!!** No GalvSkinResp if damaged.

L. Hippocampus   R. Hippocampus

L&R Amygdala CRF Punishment

L. Thalamus   R. Thalamus

Reward: L. Accumbens ⊕   ⊕ R. Accumbens: Reward
Dopamine   Dopamine

L. Cerebellum   R. Hypothal.   L. Hypothal.   R. Cerebellum

Primary Memory   **Id: Does**   Primary Memory
Thanatos: is, Hates   **Superego: IS, Loves**

L. Striatum   R. Striatum

L. Pons   R. Pons

L&R Locus Coeruleus CRF Fear ⊖ ⊖

memory data to see how the two data sets are *different and unrelated*. The rapid detection of differences is also of great survival value, for example noting the critical difference between the playful Poodle from the past, and the present rapidly approaching Pit Bull, foaming at the mouth.

The RH's wiring motif is regionally distributed, facilitating a global type of analysis. In contrast, the LH wiring architecture is of a local type to focus on the analysis of important details. The cerebral asymmetries, caused by the left local vs. a right-distributed wiring organization, lead to detectable laterality differences in how the corresponding vertical cortical mini-columns themselves are organized and interconnected in general. Thus, the presence of the two different cerebral hemisphere packages supports the necessary segregation of two incompatible brain processes into separate top-down and bottom-up data analysis systems (35, 69-71, 73, 131, 136, 137, 140).

The eye input assignments given the two hemispheres further reinforces this separation of hemispheric function. The more visual-global RH attends to the entire background spatial-visual field, while the left attends only to the right foreground details (14, 42, 118, 119). RH stroke or other injury results in left hemi-neglect, leading to drawings of clocks or flowers with numbers or petals only on the right side. In addition, the localized language centers in the LH make this hemisphere the more auditory-speech oriented of the two.

To have two such high-speed specialized data analysis systems on-board and intercommunicating with the Executive Ego has enabled mammals, especially humans, to be highly successful during the intense, ongoing process of survival. The contrasting processing motifs of the two cerebra show behavioral output differences that influence the thinking and behavioral orientation of each side of the dual brain.

Hemisity vs. Hemisphericity

Unfortunately, academic psychologists, attempting to impress the world that psychology was indeed a true science, defined hemisphericity in a hyper-quantitative but dysfunctional manner. They decided that individual human hemisphericity values should range over a quantitative gradient from the left-brained extreme to the right-brained extreme. When subjects took tests based upon that definition, almost no one marked the extremes. Instead, most of them conservatively marked themselves to be near the middle of the road. How can one separate any body into hemisphericity subgroups with such a political correct system! As an ultimate result, the whole hemisphericity field crashed and burned in the 1980s and became a stigmatized and blackballed subject (4), where for decades no grants or promotions were awarded to academic psychologists daring to do research that area.

This all happened because no quantitative methods existed to sort which, if any of the sometimes exorbitant claims for hemisphericity were true. Until this day, essentially no further research on this subject has been published. Yet, most everyone in the global literate population knows about right-brained and left-brained people. These differences between individuals obviously exist, even if academic psychology has declared hemisphericity a forbidden topic.

For reasons mentioned in the Preface, I came to recognize the inevitable, unavoidable biological origin and existence of right and left brain-oriented people. Since, I already had earned tenure permanence at the University of Hawaii, School of Medicine, and had the very small amount of personal funds needed to begin the research, I was free to investigate. Getting my results published was another matter! However, over the years 12 reports (85-94) have appeared in peer-reviewed neuroscience journals

Development of Biophysical Methods to Measure Hemisity

It was clear that good methodology to separate right and left brain-oriented individuals was lacking in the field of behavioral laterality. To make things easier, I threw away the idea that one's hemisphericity was located somewhere on a range between left to right behavioral extremes. Instead, I greatly improved my chances of developing good methods by redefining the topic in a more useful manner. I created a shorter term, called "Hemisity", defined by asserting that one was born either a right brained-oriented person, or left brain-oriented person. Then, over a period of several years, I developed five different biophysical methods that segregated a large group of people into the *same two* approximately equal sized groups (85-88, 90-92). This was promising!

Using one of the best old hemisphericity questionnaires and some published new ones of my design, we then found that group members from one of these repeatedly selected groups chose right brain-oriented answers, and that those in the other repeatedly selected group mainly chose left brain-oriented answers. Eureka! We had developed five independent quantitative biophysical methods to separate people into right and left brain-oriented groups. They were Dichotic listening (85-86), Mirror Tracing (87), Line Bisection (88), Hemisity (84), and Magnetic Resonance Imaging (MRI) (91, 92). Being biophysical, these methods were not influenced by race, education, age, language, or culture. With these I was able to categorize about 250 of my associates, friends, and relatives in terms of right or left hemisity.

Discovery of Brain Structural Differences Between Hemisity Subtypes

Further, through use of MRI, we were able to inquire if there might even be brain structural differences between right and left brainers. Thus far, we have discovered two. The first of these was that the corpus callosum cable system between the two sides of the brain was up to three times larger in right brainers than in lefts. We shall see the significance of this later.

The second neuroanatomical difference we found was in an important element of the brain's executive system (16, 27, 28, 39, 99, 132, 134) that on average was twice as thick on one side of the brain as on the other (60, 77, 81, 100, 101, 151) . Technically speaking, this location was the ventral gyrus of the anterior cingulate cortex (vgACC), the darkened structure, which includes Areas 24 and 24' of the frontal cortex in **Figure 3.**

Discovery of a Unilateral Executive Observer

Even more interesting, the thicker vgACC side oc-curred on the right or left side in subject's brains in a seemingly random or idiosyncratic pattern of sidedness (138). That is, for any individual, it was impossible to predict which side of the vgACC would be the thicker. However, if their right or left-brain hemisity group was known by other means, then the thicker side of their ex-ecutive was almost always on the same side as their hem-isity, in fact 98% of the time (92). This was good confir-mation of the validity our earlier biophysical methods (87, 88).

People with Left sided and Right sided Executives See Things Differently.

This put us in position finally to ask the most im-portant question: Did these two hemisity subtypes show different thinking and behavioral styles? We put together over a hundred "either/or" forced choice questions that we thought the two hemisity subtypes might answer dif-ferently. One question that didn't work was "which fla-vor of ice cream do you prefer? chocolate or vanilla?" Amazingly, however, for 30 of these questions there was a significant difference between answers from left brain-ers and right brainers. **Table 1** shows the results (94).

Why and how should one's thinking and behavior-al styles be affected depending whether if their executive is either on the right or on the left side of the brain? As you know, there is a deep cleft running down the middle

Figure 3. Executive Observer on One Side of the Cingulate Cor-textex.

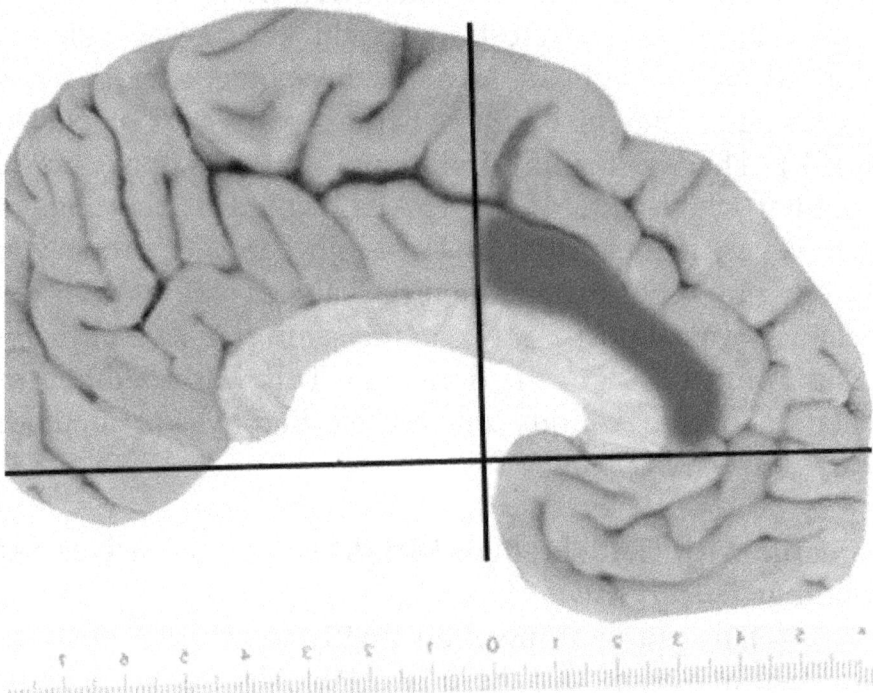

of your upper brain from front to back. The cerebral hemispheres on either side are separated completely, with only some coaxial lines of interconnection, including the variable sized corpus callosum, the lighter-colored object at the center of **Figure 3**.

As we have learned, the reason the hemispheres are separated is because their processing activities are opposite and cannot be done in the same space. In about 90% of humans, the LH is the language side of the brain, which is doing top-down analysis, ever probing deeper into the sublevels of the universe. At the same time, the RH is doing bottom-up processing, continually looking at how an element fits into a larger order at the next higher level.

The in utero embedding of the executive on one side of the brain or the other (21, 133) influences its perspective on life, and thus, is very important to a person's individual thinking and learning style. This asymmetry is lifelong (103). That is, an executive on the left looking for differences is the famous "Splitter" type of personality, who spots important details. The one on the right, looking for similarities of a global nature, is the well-known "Lumper" personality, who sees the big picture.

Table 1 confirms these ideas. The right and left hemisity individuals differed from each other in five areas: 1. Logical orientation; 2. Type of consciousness; 3. Fear level and sensitivity; 4. Social and professional orientation; and 5. Pair bonding style and Spousal dominance. In general, each question was correct for a subject's hemisity about 80% of the time. That is, 80% of right brainers were night owls, while 20% of right brainers were morning larks (20).

As you will see from **Table 1**, a few of the old right and left hemisphericity properties remain to which many more have been added.

Right or Left Brain Oriented? Find Out Your's and Other's Hemisity Subtype.

By using only **Table 1**, in most cases, it becomes quite possible to guess one's own and/or another's hemisity subtype. However, in the Appendix of this

Table 1. Thirty Behavioral Correlates of Hemisity

LEFT BRAIN-ORIENTED PERSONS	**RIGHT BRAIN-ORIENTED PERSONS**

LOGICAL ORIENTATION

Analytical (stays within the limits of the data)	Sees the big picture (projects, predicts)
Uses logic to convert objects to literal concepts	Imagines, converts concepts to contexts or metaphors
Decisions based on objective facts	Decisions based on feelings, intuition
Uses a serious approach to solving problems	Use a playful approach to solving problems
Prefers to maintain and use good old solutions	Would rather find better new solutions.

TYPE OF CONSCIOUSNESS

Daydreams are not vivid	Has vivid daydreams
Doesn't often remember dreams	Remembers dreams often.
Thinking often consists of words	Thinking often consists of mental images
Comfortable and productive with chaos	Slowed by disorder and disorganization
Can easily concentrate on many things at once	Concentrates on one thing in depth at a time
Often thinking tends to ignore surroundings	Observant and in touch with surroundings
Often an early morning person	Often a late night person

FEAR LEVEL AND SENSITIVITY

Conservative and Cautious	Bold and Innovative
Sensitive in relating to others	Intense in relating to others
Tend to avoid talking about emotional feelings	Talks about own/others feelings of emotion
Suppresses emotions as overwhelming	Seeks to experience/ express emotions
Would self-medicate with depressants	Would self-medicate with stimulants

SOCIAL AND PROFESSIONAL ORIENTATION

Independent, hidden, private, and indirect	Interdependent, open, public, and direct
Avoids seeking evaluation by others	Seeks frank feedback from others
Usually tries to avoid taking the blame	Takes blame, blames self, or apologizes
Does not praise others nor work for praise	Praises others, works for praise of others

PAIR-BONDING STYLE AND SPOUSAL DOMINANCE

After an upset with spouse, needs to be alone	After upset with spouse, needs closeness
Tolerates mate defiance in private	Difficult to tolerate mate defiance in private
Needs little physical contact with mate	Needs a lot of physical contact with mate
Tends not to be very romantic or sentimental	Tends to be very romantic and sentimental
Prefers monthly larger reassurances of love	Likes daily small assurances of mate's love
Thinks-listens quietly, keeps talk to minimum	Thinks-listens interactively, talks a lot
Does not read other people's mind very well	Good at knowing what others are thinking.
Often feels their mate talks too much	Feels mate doesn't talk or listen enough.
Lenient parent, kids tend to defy	Strict, kids obey and work for approval

book, four of our published hemisity preference questionnaires used to obtain **Table 1,** are reproduced, along with how to grade them. These will enable you to determine your and other's hemisity more accurately. As a suggestion, perhaps rather than answering them by writing in this book with a pencil, you might wish to scan several copies and use them test your family and your friends. Neither questionnaire is absolutely accurate, but in combination, they are overwhelmingly suggestive of your hemisity.

To discuss hemisity efficiently, the following nomenclature was developed: A right or left brain-oriented person is called a RP or LP. A right or left brain-oriented male is a RM or LM, while a right or left brain-oriented female becomes a RF or LF.

Table 2 compares additional interesting and logical differences between the hemisity subtypes. These have been noted anecdotally, but not yet subjected to rigorous quantification. Note that in addition to thinking and behavioral differences, as with the brain, some physical differences have also been observed **(Figure 4).**

"Men are From Mars, Women are from Venus" Tells Only Half the Story

John Grey wrote a popular book called: Men Are from Mars, Women Are from Venus: A Practical Guide for Improving Communication and Getting What You Want in Your Relationships (52). This book did quite a good job in identifying the major behavioral differences between RFs and LMs that often lead to marital misunderstanding and discord.

Table 2. Other Differences Between Hemisity Subtypes: Anecdotal

Issue:	LPs:	RPs:
PersonalityType:	High sensitivity	High intensity
Taste:	Prefer unseasoned	Prefer spicy
Smell:	Aroma sensitive	Odor insensitive
Stress:	Vulnerable	Resistant
Basal Fear:	Anxious	Bold
Dream content:	Monsters, Falling	Humiliation, Excretia
Drugs of abuse:	Relaxants (alcohol)	Stimulants (amphet.)
Immune strength:	Weak	Strong
Proneness to Illness:	Often ill	Rarely ill
Medicine side effects:	Common	Uncommon
Proneness to Obesity:	Often thin	Often overweight
Bust size:	Smaller	Larger
Non-erect penis length	Longer	Shorter (See Fig. 4)
Longevity:	Type A mortality	Youthful
Mental Health issues	Alcoholism, PTSD Pedophilia	Dyslexia

Dr. Grey had no idea that hemisity existed or, more importantly, that for each sex there are two opposite hemisity types. Due to regional population or personal bias, he assumed that women in general were represented by his mainly RF (unknown to him) patients and that all men were the same as his mainly LM subjects. For several reasons, there were lesser numbers of RMs and LFs seeking his marital counseling. This led him to over-

Figure 4, Size Contrast Between Matripolar and Patripolar Non-Erect Penis Length.

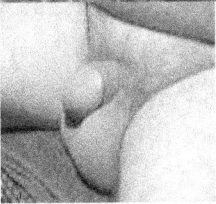

The matripolar (LM) penis (top) cannot shrink even in death. The patripolar (RM) penis (bottom)is short when not in reproductive use.

generalize his observations and to overlook the existence the very masculine RMs and very feminine LFs. When forced by questions from some of those who found his book's characterizations to be opposite of their own experience, he tended to say that such was because they were not comfortable with their sexuality and that this was confusing them.

Mars and Venus Reinterpreted

I am a RM (RxM). When I read the Mars-Venus book, the behaviors assigned to men seemed completely alien to me. Since I am somewhat daring masculine person who once set a world record for cross-country distance flown in a hang glider and who loves women, I knew something had to be wrong with the analysis in his book. Indeed, when I changed the pronouns in the book from "she" to "he" or her to him and converted all the women words to men words, and vice versa men to women, a remarkable thing happened. The book's descriptions fit me like a glove. When upset, I do not go into my cave, I need to talk things through.

The same thing happened for my second wife. She thought the Mars-Venus book was rubbish until I wrote in those changes, then she strongly identified with the many situations described as male in the book. When she was upset, she didn't need to talk, she wanted to be alone in her cave. So, we were having the same difficulties communicating as those couples in the Mars-Venus book had. However, instead of a stereotyped woman complaining about her wordless husband who never told her he loved her until his dying day, my feminine LF wife didn't like to talk, as many RFs feel compelled to do so. As a RM, I often spoke metaphorically and needed to be heard. She didn't understand metaphor talk and spoke about specifics. I wanted her affection and intimate sharing. Although she loved me, she had nothing to give, but yet needed my admiration.

Hemisity, Not Sex, is the Source of Many Behavioral Differences

As seen in **Tables 1 and 2**, RMs and RFs are more alike than RMs are to LMs or RFs are to LFs. The same goes for LMs and LFs who are also behaviorally more alike than they are to RMs and RFs. **Table 3** further supports that many so called differences between the sexes are not due to sex at all, but differences between R and L hemisity subtypes. **Table 3** lists 16 hemisity items from Table 1 that were formerly thought to be sex differences, but which turned out to be hemisity differences, independent of sex. That is, RMs carry their feelings on their shirtsleeves every bit as much RFs do. And, LFs need to go in into the silence of their caves, just as much as LF men do. This requires us to redefine carefully what the true behavioral differences between the sexes are.

Right Hemisity Persons Have Significantly Bigger Corpus Callosi than Lefts

This one reason why behavior between males and females with the same hemisity (RMs and RFs or LMs and LFs) is more similar than that between same sexes with different hemisities (RFs and LFs or RMs and LMs)

Table 3: Sixteen Behavioral Correlates of Hemisity Erroneously Tied to Sex

LEFT BRAIN-ORIENTED PERSONS *LM or LF, NOT JUST MEN!!!*	*RIGHT BRAIN-ORIENTED PERSONS* *RM or RF, NOT JUST WOMEN!!!!*
LOGICAL ORIENTATION	
Uses logic to convert objects to literal concept.	Imagines, converts concepts to metaphors
Decisions based on objective facts	Decisions based on feelings, intuition
TYPE OF CONSCIOUSNESS	
Daydreams are not vivid	Has vivid daydreams
Can easily concentrate on many things at once	Concentrates on one thing in depth at a time
FEAR LEVEL AND SENSITIVITY	
Sensitive in relating to others	Intense in relating to others
Tend to avoid talking about emotional feelings	Talks about own/others feelings of emotion
SOCIAL AND PROFESSIONAL ORIENTATION	
Does not praise others nor work for praise	Praises others, works for praise of others
PAIR-BONDING STYLE AND SPOUSAL DOMINANCE	
After an upset with spouse, needs aloneness	After upset with spouse, needs closeness
Tolerates mate defiance in private	Difficult to tolerate mate defiance in private
Needs little physical contact with mate	Needs a lot of physical contact with mate
Tends not to be very romantic or sentimental	Tends to be very romantic and sentimental
Prefers monthly larger reassurances of love	Likes daily small assurances of mate's love
Thinks-listens quietly, keeps talk to minimum	Thinks-listens interactively, talks a lot
Does not read other people's mind very well	Good at knowing what others are thinking
Often feels their mate talks too much	Feels mate doesn't talk or listen enough
Lenient parent, kids tend to defy	Strict, kids obey and work for their approval

is due to the fact that those behaviors are hemisity behaviors, not sexual behaviors. Our published MRI studies reinforced this new concept. We looked at the size of the midline cross sectional area of the corpus callosum (the light colored rotated C-like structure in **Figure 2**) (1, 108). This structure has a three-fold size variation between individuals (56, 64, 67, 74, 95, 150). Such a large

variation has invited curiosity. Many studies have unsuc-
cessfully attempted to link these differences to sex (6, 19,
58, 82, 97) or handedness (78, 113, 139, 141, 142).

We found that the corpus callosal size was significantly greater for RPs than in LPs (91). That is, RMs and
RFs could have up to three times greater interhemispheric communication than LMs and LFs did (13, 19, 53, 63,
107, 108, 127). Thus, there are actually two basic types
of men and two basic types of women. Rights of either
sex have larger corpus callosi and are behaviorally more
similar to each other than lefts of either sex with smaller
corpus callosi. This size difference had been predicted to
exist to account for the earlier reported dichotic deafness
of LPs (85, 86)

Hemisity Estimates of Historic, Scientific and Public Figures:

Once internalized, the basic understanding of hemisity makes estimating the hemisity of others quite possible. This is especially true when using biographical or
observational information. As may be seen in **Table 4**,
each of the four hemisity subtypes are represented among
the notorious composers, rulers, leaders, Nobel laureates
and performers.

Hemisity Sorting During Education and in the Professions

Over 650 students and faculty at the University of
Hawaii were assessed for their hemisity, producing the
results of **Table 5** (90). (This last table in the chapter

Table 4, Famous People within their Hemisity Subtypes:

RMs	LMs	RFs	LFs
J.S. Bach	Claude Faure	Joan of Arc	Mother Theresa
Wolfgang Mozart	P. I. Tchaikovsky	Catherine the Great	Jane Goodall
Jack Kennedy	Jimmy Carter	Oprah Winfrey	Jackie Kennedy
Bill Clinton	Al Gore	Margaret Thatcher	Laura Bush
Francis Crick	James Watson	Marie Curie	Dorothy Hodgkin
Craig Venter	Francis Collins	Barbara McClintock	Rigoberta Menchu
Sean Connery	George Bronson	Sophia Loren	Nichole Kidman
Mel Gibson	Clint Eastwood	Salma Hayek	Sandra Bullock

should be labeled **Table 5,** not Table 2 as it was when published.) While proceeding from high school on through college and graduate school and on into the professions, hemisity distributions shifted dramatically, as more hemisity sorting occurred at each stage.

That is, we found that at the high school level in the US public school system, the number of right and left brain-oriented students was almost identical. However, the sorting involved in going from high school on to college, resulted in a 7% shift. Completing college caused a 34% shift. By the time they had become Faculty, there was up to a 57% difference between the hemisity of professionals within the 18 university departments tested.

Why did increasing hemisity sorting accompany higher education or training? A probable explanation is that it resulted from each person being exactly who they were (and were not) and doing what they liked to do the best while avoiding what they did poorly. If they failed in

Table 5. Hemisity Distributions Within Populations of Fifteen Professions

Table 2: Brain hemisphericity distributions within populations of fifteen professions (n=421)

GROUP percent participation	n	LEFT BRAIN	Left Males	Left Females	RIGHT BRAIN	Right Males	Right Females
Unsorted College Entrants	228						
Western Civilization students 62	228	57%	19%	38%	43%	22%	21%
Specialist Populations	422						
Microbiology Professors 74	14	86%*	72%	14%	14%	14%	0%
Biochemistry Professors 95	18	83%*	72%	11%	17%	17%	0%
Physics (particle)Professors 80	15	73%	73%	0%	27%	27%	0%
Philosophy Professors 73	11	73%	54%	19%	27%	27%	0%
Mathematics Professors 93	27	70%	70%	0%	30%	30%	0%
Accountancy Professors 75	9	67%	44%	22%	33%	22%	12%
Law Professors 83	19	63%	32%	31%	37%	21%	16%
Art Professors (vs. Artists) 92	27	63%	38%	25%	37%	29%	8%
Civil Engineering Professors 89	17	53%	53%	0%	47%	41%	6%
Clin. Psychologists (yel. pages) 75	29	52%	24%	28%	48%	28%	20%
Electrical Engineering Profs. 75	16	50%	50%	0%	50%	44%	6%
Physicians (Medical Students) 80	178	49%	25%	24%	51%	26%	25%
Mechanical Engineering Profs. 75	9	44%	33%	11%	56%	56%	0%
Architecture Professors 100	12	33%*	26%	4%	67%	61%	9%
Astronomy Professors 66	21	29%*	30%	0%	71%	60%	10%

* $p < 0.05$.

(yel. pages) = American Psychological Society Members advertising in the yellow pages of the Honolulu phone directory.

(Medical Students): due to extremely low attrition rates of medical students, it was convenient to test them in mass rather than scheduling a separate appointment with each of them after they became clinicians.

one course and excelled in another, they could hardly be blamed for proceeding in the direction of their success. Thus, often career specializations tend to match a person's right or left brain-orientations.

As apparent from **Table 5(2)**, some left brain-oriented professions (with a top-down view of components at lower universe levels) include Particle Physics, Microbiology, Biochemistry, Mathematics, and Accountancy. Some right brain-oriented professions (using a bottom-up assembly of systems at the next higher universe level) include Astronomy, Architecture, Mechanical Engineering, Art, and Music.

Hemisity vs. Sex in the Workplace

In the Dual Quadbrain Model, we each have a "double-brain", which operates by a "Double Standard" (Neuroreality, 93). Our left brain competes against aliens as enemies using win-lose ethics. In contrast, our right brain cooperates with family members using win-win ethics. Thus, there will be two polarity styles of interaction: LPs often form bottom-up competitive democratic hierarchies, like monkey islands. In contrast, RPs work with loyalty for their often autocratic leader through top-down chains of command. Polarity style conflicts can occur in "mixed groups" due to these opposite modes of thinking. In left brain-oriented professions, dominance hierarchies may form to avoid the natural tendency for RPs to lead.

The complaint by RFs that an impossible-to-penetrate "Glass Ceiling" exists in the business world, can definitely apply to RMs as well. Importantly, it rarely exists for LFs, older feminist doctrine notwithstanding. Another part of the explanation is that only RMs and RFs have been found to be dyslexic, and that these may reach their "level of incompetence" as a glass ceiling. Non-dyslexic RPs are truly formidable males and females who have often been world leaders.

Many traits commonly considered sexual characteristics are actually hemisity traits. LPs, male or female, tend to be somewhat muted, quiet, impersonal, anxious, and emotion avoidant. In contrast, RPs, male or female, tend to be more charismatic, intense, talkative, and bold. On television, such shows as "Switching Families". unwittingly stereotype these reverse familial polarities.

Compared to lefts, rights are more likely to use touch to convey feelings of closeness; these feelings could be sexual in nature, but not necessarily. Interestingly, rights are more likely to exercise power strategies than lefts. Compared with lefts, rights of both sexes are more likely to engage in manipulative behavior
and to exercise negative or confrontational conflict behaviors. Finally, rights are more likely to enact self-disclosure behaviors, and use task-sharing in an effort to maintain their relationship.

Visible Behavioral Differences Between Hemisity Subtypes

RMs and RFs are publically charismatic, self-confident, innovative, concerned with the big picture, transparent, and inclusive with a cooperative win-win friendly attitude. In terms of religion, their experience is personal, profound and emotional (93). RMs appear to have founded all of the world religions (Chapter 10).

In contrast, LMs and LFs are publically reserved, anxious, conservative, concerned with the important details, opaque, exclusive with a competitive win-lose adversarial attitude. Their religious experience is more non-emotional, legal and abstract, including many high-church, symbolic artistic representations and rituals. The knowledgeable, even when observing strangers, can spot RP vs. LP differences.

Remember: Hemisity is Different From to Handedness.

Handedness has nothing to do with hemisity. This point cannot be over emphasized!

CHAPTER 2. Evidence Part 2: FIND YOUR FAMILIAL POLARITY

How the Existence of Hemisity Provides Direct Evidence That There Are Two Human Species.

The idea that two pre-racial biologically different "true-breeding" human species presently exist, requires a paradigm shift in thinking. It opens the door to many new possibilities and provides clarity to the nature of problems currently not understood and that are leading to multileveled human conflict and suffering. The evidence supporting the existence of these two species, which are found within *all* races, will continue to be developed in this chapter.

Observation #1: The Hemisity of Spouses: Opposites Attract

After testing the hemisity of thousands of students and faculty within the University of Hawaii, the next question that arose was: What might the hemisity of members of heterosexual couples be? Are like-like hemisity partners more attracted to each other? Or, does the old saying "Opposites Attract" actually apply to the hemisity pair bonding of spouses?

I looked at the 412 partners of 206 couples who had been together longer than 5 years (84). If you think about it, you will notice that there are only four possible hemisity couple combinations. Two are complementary:

RM-LF, RF-LM, and the two are the same-same matches: LM-LF, and RM-RF. The distributions found in this unpublished study were as follows: Of the 412 heterosexual partners, by definition 50% were men and 50% were women. In terms of hemisity, there were 53% LPs and 47% RPs, a fairly common distribution. Among the four possible partner combinations,

40% were LM-RF pairs,
27% were RM-LF pairs,
20% were LM-LF pairs, and
13% were RM-RF pairs.

Thus, 33% (20+13) of the pairs were between partners of like hemisity. In contrast, *twice* as many: 67% (40+27) of the pairs were between partners of opposite (complementary) hemisity.

Observation #2: Right-Brained Spousal Dominance Within the Four Possible Hemisity Couple Types:

Among the complementary couples (RM-LF or RF-LM), the right brain-oriented partner (male or female) was usually the *de facto* leader of the nuclear family. **Table 6** lists six paired elements from the list of 30 pairs of differences between LPs and RPs from Table 1 (94) of chapter 1 supporting RP spousal dominance in the home. From these it may be seen that in the RM-LF or LM-RF pairs, the RP was dominant, male or female, as the case may be.

However, in the same-same RM-RF pairs, leadership was hotly contested, leading to a relative instability of this pair. In contrast, the LM-LF pairs were the most

Table 6. Six Spousal Dominance Oriented Items

LEFT BRAIN	PAIR-BONDING STYLE*	RIGHT BRAIN
Does not read other people's mind very well		Very good at knowing what others are thinking
Avoids talking about their own & other's emotions		Often talks about their and other's emotions
Can tolerate it if their mate defies them in private		Finds it intolerable if mate defies them in private
Likes longer-term, larger rewards of mate's love		Likes daily small reassurances of mate's love
Often feels mate talks too much		Often feels that mate doesn't talk / listen enough
Not a very strict parent, kids tend to defy		Strict, kids obey and work for his / her approval

*from Table 1, Chapter 1

stable of the four. There, leadership usually fell on the larger (often-unwilling) male. Thus, the relative stability of the possible hemisity couple combinations appears to be: LM-LF > RM-LM and LM-RF > RM-RF. Recall, that the relative stability of the hetero and homo sexual couples has seemed to be: F-F > M-F > M-M.

Marital dominance is not to be confused with work dominance outside of the home. As mentioned in Chapter 1, at work LPs often form highly competitive exclusive "It's who you know" groups in an unconscious attempt to avoid RP authoritarianism.

Observation #3: The Hemisity of Children from the Four Hemisity Couples: Fixed vs. Random

The question that arose was: Might the hemisity of the children from each of the four couples differ, and if so, in what way? **Figure 5** illustrates findings on off-spring hemisity from the four possible hemisity couples. These were based upon 3-5 generation genealogies from

14 unrelated families (84). Significantly, for the complimentary couples, both the RM-LF pairs, and the LM-RF pairs, hemisity was usually "like father, like son", and "like mother, like daughter". In contrast for the RM-RF couples, the children's hemisities were random and unrelated to that of the parent of same sex. In a later study, the offspring of LM-LF couples was found to be mostly LPs.

Simplest Explanation for the Data Thus Far: Two Unrecognized True Breeding Human Species Exist

Based upon these and later observations, it was concluded that in nature, "Opposites attract" refers to the hemisity of the wild type reproductive pairs of two different true-breeding human species. And unlike the children of either true-breeding lineage, the children of the like-like cross-breeding couples are hybrids between these two existing human species. They do not breed true and their children are cross-wired, as we shall see. These two newly recognized human species here are called *Homo sapiens patripolaris* and *Homo sapiens matripolaris*. Members of these separate species will be referred to as patripolars and matripolars.

Further, it is asserted that these two lineages utilize two equally possible, but opposite reproductive strategies (41, 80, 83, 111, 148) here identified under the new term: "Familial Polarity". These are the "patripolar" (RM-LF) and "matripolar"(RF-LM) reproductive strategies, both of which were firmly established long before the vertebrates emerged (see Table 13).

HEMISITY GENEAOLOGY: TWO TRUE- BREEDS MAINTAINED

PATRIPOLARS: (RM-LF)

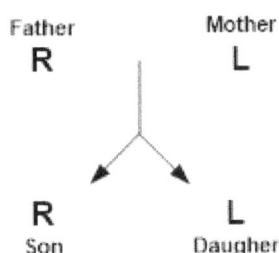

Father
R

Mother
L

R
Son

L
Daugher

MATRIPOLARS: (RF-LM)

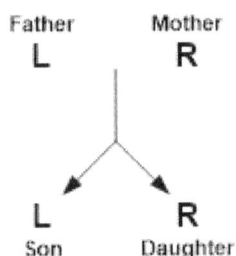

Father
L

Mother
R

L
Son

R
Daughter

SONS: SAME HEMISITY AS FATHERS
DAUGHTERS: SAME HEMISITY AS MOTHERS

HEMISITY GENEAOLOGY: HYBRIDS FORMED

L-L HYBRID PAIRS

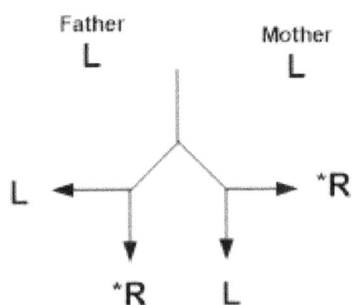

Father
L

Mother
L

L
*R

*R   L

SONS AND DAUGHTERS:
MOSTLY LPs

* Miscarriage loss of Rs

R-R HYBRID PAIRS

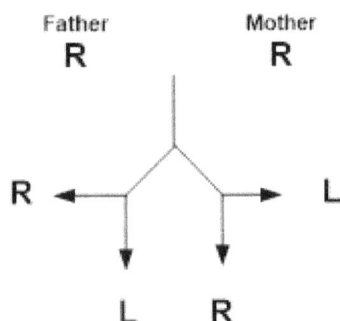

Father
R

Mother
R

R
L

L   R

SONS AND DAUGHTERS:
RANDOM HEMISITY

41

Observation #4: Hemisity and Corpus Callosum Size

As mentioned in the preceding chapter, the corpus callosum is one structure of the brain that manifests an up to three-fold size variation between individuals. This structure is the bundled information bridge between the isolated right and left sides of the cerebrum. In **Figure 3**, the corpus callosum can be recognized as that lightest colored object at center of photo of the brain that looks like a letter C rotated clockwise. As mentioned in Chapter 1, people have attempted to tie this variation in the midline corpus callosal cross sectional area with sex or handedness without success.

Once we discovered and could identify individual hemisity, we asked whether there was any relationship of corpus size to hemisity subtype. **Figure 6** shows the results of our MRI study (92). There, it may be seen that indeed the corpus callosal area of RPs was significantly larger than that of the LPs. Further, the relative size relationships between the two sexes for both the patripolar and matripolar couples were the same. In both pairs, the LPs with smaller corpus callosi were significantly less interconnected between their two brain hemispheres than the RPs were. This repeated pattern is consistent with what would be predicted if they were the two familial polarity human species.

Clearly we are not dealing with differences in IQ here, since representative members of each of the four hemisity classes included some tenured faculty at the

Figure 6: Comparison of Corpus Callosal Size With Hemi-sity Subtype

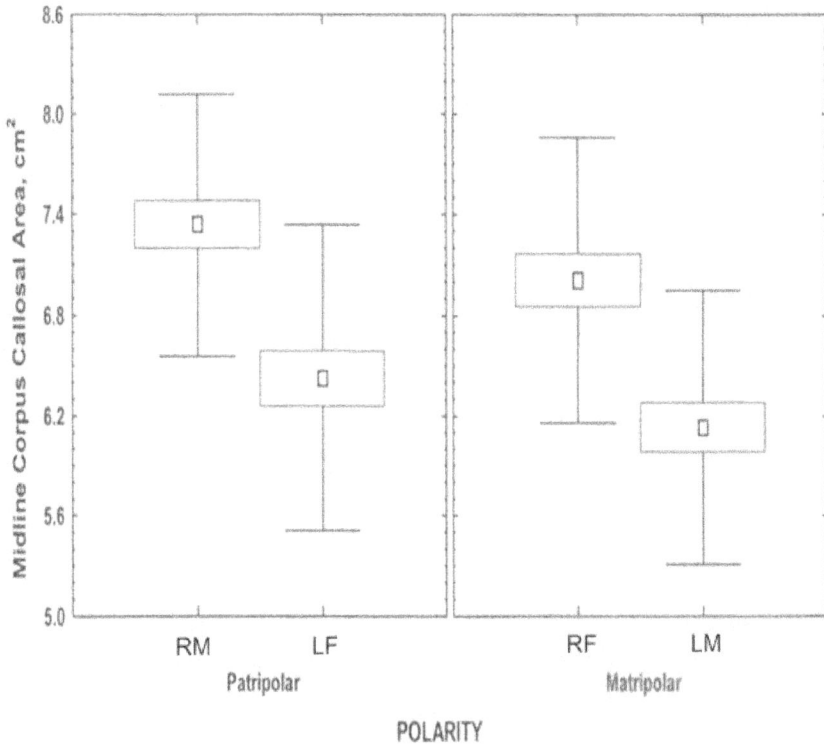

university. In this sample of over a hundred persons, the smallest sized corpus was found in a popular university dean. His wife, a RF humanities professor, by chance, had the largest one.

Observation #5: Evidence that Hemisity is Inherited:

Twin studies clearly show that corpus callosum size is inherited (129). Also, if all these hundred subjects of Figure 2 were members of the same species, then one would need to combine the RMs and LMs into a huge corpus callosum size smear. The same would occur if both females were mixed together. Hemisity sorting made the peaks much narrower.

The above genealogical studies also suggest that the hemisity of offspring follows exact inheritance rules (25). These and other anatomical and evidence suggest that, in general, one's hemisity is anatomically laid down during pregnancy and is unchangeable.

Observation #6: Opposite Reproductive Strategies: Foundation of Endless Conflict

Patripolar Reproduction (RM-LF): Among the gorillas (46) and orangutans (49) in the wild, the patripolar haremic strategy of *male dominance* is as follows. The winner of the ongoing battles between males strongly attracts females, who are always looking for the best sire to father their children. He thus becomes the leader of a harem. The monogamous females of the harem thus formed, strongly reject the male losers. Therefore, the ostracized defeated males tend to form isolated bachelor groups. There, the males grow and practice their martial arts, awaiting the opportunity to overthrow the current alpha male, and thereby gaining reproductive access to their own genetic immortality.

Thus, in patripolar lineages, paternity of offspring usually traces to the alpha male. Because, size significantly contributes to winning, there is often a sexual dimorphism in patripolar pairs, where the male can be much larger than, even more than twice the size of the female. Male dominant patripolar groups are generally non-territorial. Instead, they are invasive of all territory and often migratory.

Matripolar Reproduction (RF-LM): In the perhaps less appreciated matripolar, orgeic (orgy having) reproductive strategy of the Chimpanzees (57) and Bonobos (29), the males *compete for territory* instead of for leadership. For example, they seek to claim the property next to an important resource, such as a water hole, as being the most desirable territory. The least valued real estate lies on the outskirts. Even some insect males sort and segregate themselves into this geographic hierarchy of territory, as for example in so-called lek mating.

Most importantly, in the matripolar reproductive strategy, the male welcomes any polygamous female that enters his territory. If in estrous, she is repeatedly inseminated as she passes through his and his competitors territories on her way to the water hole. Thus, in the matripolar reproductive strategy, the males make love not war. Paternity is usually quite random, and each male supports and protects any mother with child as if it were his very own.

Observation #7. Survival of the Fittest Competition Occurs at Different Universe Levels in the Two Species

As indicated in **Table 7**, clearly, the two reproductive strategies differ regarding at which universe level the

Table 7. Two Reproductive Strategies Result in Opposite Familial Polarities

Variable	Patripolar (RM-LF)	Matripolar (RF-LM)
Dominant Sex	Male	Female
Promiscuous Partner	Male	Female
Paternity	Alpha Male	Random
Males must fight for reproductive access	Yes	No, but are territorial
Males are much larger than Females	Yes	No
Universe level of competition	Organisms: Males Battle!	Cells: Sperm Races!
Larger sex organs required to compete	No	Yes

killing violence of the "survival of the fittest" occurs. In the patripolar reproductive strategy, the competition occurs at the organism level between battling males. Females follow, court, and are inseminated by the winner of these fights. Because no competition within the female reproductive tract exists, penises and testicles are relatively small in patripolar males.

By remarkable contrast, in the matripolar reproductive strategy, the killing competition occurs at the cellular level. When each of the females comes into heat, there is a big party. Essentially all matripolar males take turns copulating with her. This ends in a great race among the sperm from the many semen donors within the reproductive tract of that female. Billions of sperm race to be the first to penetrate her receptive ovum. After the "one in a billion" winning spermatozoon is welcomed inside, the ovum closes her gates. The vast number of losing sperm competitors die (of frustration) over the next few days. Thus in matripolars, overall male body size is much less important and sexual dimorphism is low, or even reversed. Instead, the ability to deliver increased sperm numbers (larger testicles) with longer, more penetrative insemination systems (longer penis) has evolved, along with seminal agglutinating agents and "killer" sperm.

Today, idealized young-adult patripolar, RM human males are estimated to weigh about 200 lbs while patripolar LF females weigh something over 100 lbs. Idealized matripolar, LM, human males might weigh around 170 lbs and their matripolar, RF females around 135 lbs. While the erect penis sizes between patripolar and matripolar males are comparable overall, the *non-erect* penis length of patripolar males tends to be much smaller (as definitively illustrated in **Figure 4** of the last chapter). Similarly, on average,.non-pregnant patripolar females tend to have smaller breasts than matripolar females.

Observation #8: Opposite Courtship and Child Rearing Behaviors Exist

Table 8 summarizes and compares the currently unrecognized but hauntingly familiar biology-driven opposite behavioral dyads within Familial Polarity courtship and parenting styles. In the patripolar home, the dominant father is a charismatic right-brained big picture husband with a larger corpus callosum. He is assisted by a quiet left brain, important details, supportive wife with a smaller corpus callosum. As the role model, he sets standards for their children to earn his conditional love. His mate gives them unconditional love and prevents paternal excess.

Table 8 (not 4):

Table 4: Personality Traits within the Two Polarity Family Types

TRAIT:	PATRIPOLAR FAMILIES		MATRIPOLAR FAMILIES	
Parental Sex **Hemisity** **Corp. Callos. Size**	Male Right Larger	Female Left Smaller	*Female* *Right* *Larger*	*Male* *Left* *Smaller*
Mental Orientation **Verbosity, Speech** **Family Leadership**	Big Picture Charismatic Most dominant	Important Details Quiet, articulate Most supportive	*Big Picture* *Charasmatic* *Most dominant*	*Important Details* *Quiet, articulate* *Most supportive*
Parental Love Type **Parental Function** **Child's Hemisity** **Parental Status**	Conditional Sets standards Boys are Rights Role model	Unconditional Prevents excess Girls are Lefts Serves the child	*Conditional* *Sets standards* *Girls are Rights* *Role model*	*Unconditional* *Prevents excess* *Boys are Lefts* *Serves the child*
Mating Behavior **Mating Target**	Males select displaying females who are: Healthy, intelligent, humorous, loyal, devoted, and want to serve him.	Females court winning males who are: Tall, dark, and handsome, champs winners, strongest. Most socially powerful, richest, smartest of crop	Females select displaying males who are: Healthy, intelligent, humorous, loyal, devoted, and want to serve her.	Males court winning females who are: the most physically attractive: leanest, big-breasted. Most socially powerful, richest, and smartest of crop.

It is quite the opposite in the matripolar home. There, the mother is a dominant, charismatic right brained big picture wife with a larger corpus callosum. She is assisted by a quiet left brain, important details, supportive husband with a smaller corpus callosum. As the role model, she sets standards for their children to earn her conditional love. Her mate gives them unconditional love and protects them from maternal excess.

In terms of mating behavior, patripolar male "winners", placed upon a stage as heroes, select from among crowds of swooning patripolar female groupies at their feet as to who will best serve them.

However, no self-respecting matripolar female would think of wandering down there. Instead, these women take to the stage themselves as beauty stars. She then chooses whom among the displaying males panting at her feet would serve her best. The infamous "wardrobe malfunction", bare breast briefly exposed on television at the half time of an American Super Bowl football game a few years ago, thus caused opposite reactions of offended horror or of amusement, depending upon the familial polarity of the viewer.

Further, mature RPs of either polarity tend to be more obese than the thinner LPs. Thus, the matripolar couple stereotype of "Jack Sprat (the skinny husband who) could eat no fat, and his (heavy) wife who could eat no lean. Together they licked the platter clean."

In addition, at least among "Caucasians" the pupil-to-eyebrow vertical distance tends to be smaller in RMs, who also may have a heavier overlying brow ridge than LMs. The latter tend to have higher eyebrows and negligible brow ridges, perhaps reflecting ancient Neanderthal and Cro-Magnon ancestral differences. This phenomenon is sometimes visible on US television between Republican and Democrat news commentators.

The DNA-based genetic content of humans differs from the four apes by only 1-3%. It is a stunning fact that variations between human races are even higher (3-5%), and between individuals, this can differ by an amazing

30-50%. Thus, it is not surprising that discovery of hemisity has uncovered evidence for two distinct human familial polarity species. At present, the existence of two opposite primate reproductive strategies has yet to be recognized in apes by contemporary science, much less in humans. It is ironic that hemisity and familial polarity were also unknown to the writers of the Bible, having apparently escaped the eye of God. The Genesis story of the rejected farmer Cain, killing his brother Abel the rancher, both sons of Adam, suggests that the Bible was written from a completely patripolar perspective.

Observation #9: Existence of Contrasting Behavioral, Cultural, and Institutional Differences between the Human Polarities

In **Table 9,** the contrasting behavioral orientations and ecological niches between the patripolar and matripolar humans are illustrated. Patripolar male big-game-killing skills, animal husbandry, nomadic, inventive, and meat and potato styles are very different from those of the matripolars. The latter preferred the crafts of the non-killing gardener, vegetarian, settled stable cooperator cultures, where the arts could incubate and develop.

The ethnic database of 1170 cultures compiled by Murdoch, scanned for 15 social variables by DeMeo (26) forms the basis for **Table 10.** Obviously, the members of two polarities are quite different in their basic attitudes, behaviors, and social institutions regarding the treatment of children, sexuality, and rights of women. The male-dominant patripolars are harsher, more rigid in attitude,

Table 9. Patri-and Matripolar Adaptations and Ecological Orientations:

	Patripolar: (RM-LF)	Matripolar: (RF-LM)
Field Specializations:	Early Big-Game Hunters	Gatherers
	Midperiod Herdsmen	Gardeners
	Later Ranchers	Farmers
Species domestication:	Horse, Wheat, Barley	Ox?, Legumes?
Dietary Orientation:	Meat, Blood, Dairy	Vegetarian, Spices
Mobility:	Nomadic, Marginal Land	Stationary, Best land
Group Discipline to Survive:	Yes, Militaristic	No, Laissez faire
Society Types:	Dominator cultures	Cooperator culture
Early Arts Devel:	Barbarian Invaders	High Arts-Cultures

possibly to protect their wife and children from their fellow male's tendencies toward violence toward others. In addition, patripolar nomadism demanded a much tighter familial control in terms of the use of time and resources in order to survive. This was not a child or female indulgent situation, and required much higher discipline and order to survive in the badlands than farmers at the oases required. Contrasting comments could be made regarding matripolar gathering, gardening, farming and settlement on rich agricultural land which are inherently much more time consuming and relaxed.

Table 11 highlights similar differences in the social and religious institutions. The contrasts are dramatic and edifying. Through Familial Polarity we can begin to understand the origins of the wide diversity of human societies and cultures all around us.

Table 10. Contrasting Behaviors, Attitudes, and Social Institutions:

15 variable correlations within Murdoch's ethnic database of 1170 cultures (26)

.	Patripolar (RM-LF):	Matripolar (RF-LM): .
Infants & Children:	Less indulgence	More indulgence
	Less physical affection	More physical affection
	Infants traumatized	Infants not traumatized
	Painful Rites of Passage	Absence of pain in initiation
	Dominated by family	Children's democracies
	Sex-segregated houses	Mixed sex children's houses
Sexuality:	Restrictive, anxious view	Permissive, pleasurable attitude
	Genital mutilations	Absence of genital mutilations
	Female virginity enforced	No female virginity taboo
	Vaginal intercourse taboo	No vaginal intercourse taboos
	Adolescent sex censured	Adolescent sex freely permitted
	Homosexual-Incest taboos	Absence of Homo-Incest tendency
	Concubinage / prostitution	Absence of concubinage / prostitution
Women:	Limits on freedom	More freedom
	Inferior status	Equal status
	Vaginal bleeding taboos	No vaginal blood taboos
	Cannot choose own mate	Can choose own mate
	Cannot divorce at will	Can divorce at will
	Males control fertility	Females control fertility
	Reproduction denigrated	Reproduction celebrated

Figure 7 is a humorous and exaggerated stereotypic comparison of the two Familial Polarities today. Individuals from all four subtypes of familial polarity can be highly intelligent, and representatives of each have received Nobel prizes for scientific research.

Table 11: Contrasting Social and Religious Institutions:
15 variable correlations within Murdoch's ethnic database of 1170 cultures (26)

	Patripolar (RM-LF):	Matripolar (RF-LM:
Culture, Family	Patrilineal descent	Matrilineal descent
And Social	Patrilocal marital home	Matrilocal marital home
Structure:	Compulsive monogamy	Noncompulsive monogamy
	Often polygamous	Rarely polygamous
	Authoritarian	Democratic
	Hierarchal	Elegantarian
	Political/Econ. Centralism	Work-democratic
	Military specialists/caste	No full time military
	Violent, sadistic	Nonviolent, sadism absent
Ancient	Male/Father oriented	Female/Mother oriented
Religion:	Asceticism, avoidance	Pleasure welcomed and
	of pleasure, pain-seeking.	institutionalized.
	Inhibition, fear of nature	Spontaneity, nature worship
	Male shamans, healers	Male or Female sha/healers
	Strict behavioral codes	Absence of strict codes

Although the term, Familial Polarity, is not yet part of our cultural consciousness or vocabulary, its existence is easy to detect in the hemisity of public figures, both in politics and in the entertainment world. With a sound understanding of Familial Polarity and a careful assessment of relevant biographic material, it becomes quite possible to assess the right or left brain-orientation of historic figures as well **(Table 4)**.

Table 12 contains an estimation of the Familial Polarity of several families who became familiar to the public through the mass media.

Figure 7. The Two Polarities: Matripolar above and Patripolar

The chemistry's right: it was love at first sight for Richard Burton when he met Elizabeth Taylor.

Observation #10. Anatomical Differences Between the Two Species Exist

Figure 8 compares a matripolar Cro-Magnon skull with a patripolar Neanderthal skull. Photos of RF French actress, Bridgit Bardot and LF American actress Jane Fonda are shown beneath. The similarities of the modern individuals to these ancient exemplars are striking. Clearly the incredible Neanderthals with their huge brains can no longer be considered primitive in contrast to their Cro-Magnon contemporaries. The neuroanatomical differences, mentioned elsewhere in this book must be added to these.

At this point, you should be able to identify your own hemisity and that of some of those around you. In addition, you can begin to notice the four marital type pairs, including those of your own and of your parents.

Further Ideas on the Origin of Two Human Species.

Observation #11: Terrestrial Life Has Originated More Than Once

According to "Recreating the Universe" in the Introduction, cellular life spontaneously continually emerges within galactic halos of life. Thus, there is no reason to say this couldn't have happened more than once here on earth. In fact, at the level of the simplest life forms, modern bacteriologist taxonomists have already identified three apparently unrelated and independent superdomain families of life: the Eubacteria (monocellular), Archaea

Table 12, Patripolar and Matripolar Public Families in Daily Life

Male Dominant Patripolar Families: **Identities:**

Bush: George W and Laura	former US President and Wife
Clinton: Bill and Hillary	former US President and Wife
Reagan: Ron and Nancy	former US President and Wife
Kennedy: Jack and Jackie	former US President and Wife
Nixon: Richard and Patricia	former US President and Wife
Irwin: Steve and Terry series	*Crocodile Hunter,* 2000s TV adventure
Fonda: Henry, Peter, Jane	A family of major Hollywood movie stars
Bunker: Archie and Edith	*All in the Family"* 1970s TV comedy

Female Dominant Matripolar Families:

Gore: Al and Tipper	former US Vice President and Wife
Carter: Jimmy and Rosalyn	former US President and Wife
Thatcher: Margaret and Denis	former English Prime Minister / Husband
Meir: Golda and Morris	former Israeli Prime Minister / Husband
Ghandi: Indira and Feroz	former Indian Prime Minister / Husband
Curie: Marie and Pierre	former French Scientists
Arnaz: Ricky and Lucy	*I Love Lucy* 1950s TV comedy series
Jefferson: George and Louise	*The* Jeffersons 1980s TV comedy series
Bundy: Al and Peggy	*Married with Children,* 1990s TV series

(mono-cellular), and Eukarya (found in multicellular organisms). The five supergroups in the Eukarya are Unikonta, Excavata, Chromalveolata, Rhizaria, and Archaeplastida.

Terrestrial life is also now said to have independently originated in at least two, and possibly three locations. The first was in a warm pond at the surface of thocean; the second, near volcanic vents at the bottom of the sea; and third from inseminations of spores from outer space. Thus, for there to be more than one tree of life is not is incompatible with current knowledge.

Observation #12: There Appears To Be At Least Two Trees of Life

Thus, <u>Homo sapiens patripolaris</u> and <u>Homo sapiens matripolaris</u> appear to be branches from two different parallel trees of life. In the patripolar tree of life, the reproductive strategy is harem forming (haremic) throughout. The males compete by fighting and the females love the winner. Paternity is usually that of the alpha male.

The second tree of life with an opposite reproductive strategy is territorial and orgy having (orgeic). The males battle for territory, each one receiving a territory in rank order of perceived land value, the best being equivalent to the "nearest the water hole". Females in heat encourage any male to breed with them as they pass through the territories with paternity being random.

What is interesting is that within the ecological niches, there are increasingly complex competing forms from each of the two trees of life. This is illustrated briefly in **Table 13** and here called Dyadic Evolution.

Thus at some level, wolves and foxes should not be genetically related, nor rats and mice, nor horses and donkeys, crocodiles and alligators, and so on. As our understanding of human and animal genomes progresses, we should not only be able to spot unique DNA <u>Homo sapiens patripolaris</u> and <u>matripolaris</u> sequences, but also hybrids between these species of the two trees of life, and in utopia make necessary uncrossing corrections or repairs.

Figure 8, Matripolar Cro-Magnon and Patripolar Neanderthal Skulls and the Resemblance of Possible Female Modern Counterparts

Table 13. Brief Diagram of Potential Dyadic Evolution Stages Leading to Humans

Presently Unrecognized Omnipresent Parallel Paths:

*1. Origin of Cellular Life   Sea vent **vs.** Warm pond: Space spore?
2. Competing Chemistrys   RNA World **vs.** DNA World
3. Oldest recognized forms   Arche-bacteria **vs.** Eu-bacteria
4. Stationary and Mobile forms   Unicellular **vs.** Multicellular
5. Plants vs. Animals   lf-sufficient **vs.** Parasitic
6. Vertebrates vs. Invertebrates   Vascular **vs.** Nonvascular
7. Sexual vs. Asexual   Ovulate **vs.** Sprout
8. Fertilization   Internal **vs.** External
9. Aquatic   Sharks **vs.** Fish
10. Life moves onto dry land   Reptiles **vs.** Plants

Two Reproductive Strategies, *Haremic vs.Territorial*, existed by then:
Patripolar vs. *Matripolar*

11. Reptiles   Alligators **vs.** Crocodiles
12. Dinosaurs   Anchiceratops **vs.** Tyrannosaurus (female bigger)
13. Birds   Chickens **vs.** Peafowl
14. Lay eggs vs. Nurse infants   Reptiles **vs.** Mammals
15. Monotremes   Duck-billed platypus **vs.** Spiny Ant-eater
16. Marsupials   Red Kangaroo **vs.** Northern Quoll
17. Placentals   Deer, Antelope, Wolf, Rat **vs.** Mouse, Fox, Sheep, Elk
                Elephant, Walrus, Horse **vs.** Donkey, Hyena, Bear
18. Early Primates   Brown Lemur **vs.** Ring-tailed Lemur
19. Apes   Gorillas, Orangutans **vs.** Chimps, Bonobos
20. Hominids 7MyrBP   Neand., Peking Man **vs.** CroMagon, Narmada Man

---

* items 1-10 are only suggestive regarding Familial Polarities.

Observation #13: Two Opposite Reproductive Strategies Exist in the Apes, Our Closest Living Primate Relatives

Each of the four Ape species exhibits one of the only two sexual strategies of familial polarity. Both gorillas and orangutans use the patripolar strategy where a massive, violent, harem-forming male maintains exclusive breeding rights by fighting at the organism level. For gorillas, the males weigh around 375 lbs, while females weigh about 200 lbs. In the case of orangutans, the males weigh 225 lbs, more than twice that of the about 100-lb females. Because of lack of competition within the monogamous female tract, patripolar apes need only small penises (1-2"), and small testes, the latter of which are even internal in the gorilla.

In contrast, chimpanzees and the lesser known but amazing bonobos use the matripolar reproductive strategy. In both of these matripolar species, each time a female comes into heat, a big party occurs in the form of a sexual orgy. During this short time she welcomes all males of the troupe to make love with her (sometimes having more than 50 separate trysts in one day) while all their sperm make war inside her. Chimpanzee and bonobo males weigh about 100 lbs and their females about 80 lbs. In contrast to the patripolar males, the matripolar ape males have huge testicles and long penises (3-5") to produce and deliver overwhelming numbers of sperm to the competition.

Stages of Primate Evolution Within Each of the Two Trees of Life.

Table 14 summarizes the increases in brain size from the apes (109), on through the hominids, to early and later humans, as documented by many archeologists. It shows that a skull found in Chad, Africa, dated around 7-6 million years ago, had features intermediate between ape and human. This is consistent with Darwin's hypothesis on the "Descent of Man" (23b). Parenthically, although Darwin cites many examples consistent with familial polarity, he never recognized it, perhaps due to his basically patripolar orientation.

Other finds of more recent skulls (**Table 14**) show the cranial capacity increasing more and more, until by the time of the Neanderthals and Cro-Magnons, it exceeded that of modern humans by over 30%! In humans brain size appears to correlate with IQ (126, 129), suggesting that our ancestors were formidable survivors!

Observation 14: Throughout the Archeological Record, Matri and Patripolar Species of Hominids Existed in Parallel

Among the hominoid remains at each time period (5, 22, 57, 106) are almost always two different contemporaneous hominid lineages, one, called *robustus*, showed great sexual dimorphism, while the other, called *gracilis*, was more graceful and less stocky with little sexual dimorphism evident. This strongly suggests that both the patripolar and matripolar lineages have come forward parallel in time, for example in the form of Neanderthal brutes (120, 130) and Cro-Magnon beautiful

Table 14: Growth of Brain Size with the Evolution of Culture

Species	Location	Cranial Capacity	Existence time, yrs
Chimpanzee	Africa	321 ml	Endangered
Bonobo	Africa	336 ml	Endangered
Gorilla	Africa	425 ml	Endangered
Orangutan(African fossils)	Indonesia	443 ml	Endangered
Sahelanthropus tchadensis	Africa ("Toumai")	380 ml	6-7 million
Homo habilis	Caucasus (Dmanisi)	700 ml	2.0 – 1.5 million
Homo erectus	Asia (Java, Peking)	995 ml	1.9 – "100,000"*
Homo heidelbergensis	Europe	1300 ml	800,000-200,000
Homo neanderthalensis	Europe	1700 ml	400,000-"20,000"*
Homo rhodesienses (CroMagnon)	Europe	1600 ml	600,000-200,000
Early Homo sapiens	"Out of Africa"	450 ml	160,000 - present
Modern Homo sapiens	Global	1250 ml	present: 6 billion+

*= possibly not gone

people. True-breeding helped keep these lineages separate.

However, the proposal that ancient interbreeding between them occasionally occurred has been supported by archaeological evidence and has become a relevant topic among anthropologists. It was said that Neanderthals were primitive because they were always fighting, as demonstrated by the fact that the skeletons of many of the males have healed broken bones. Very few healed bones were found among the supposedly more peaceful

Cro-Magnons, so they were assumed to have been much smarter. More likely, due to their patripolar reproductive style, Neanderthal males still fought for leadership and reproductive access, while the matripolar Cro-Magnon reproductive style to "make love, not war" was less personally dangerous.

It is likely that with their huge brains, both lineages were more intelligent than we are. We do not have to be particularly intelligent to survive these days within modern bedroom cultures, unlike those hominids who lived through the ice ages among the glaciers, caves, or deserts.

Source of Both Patripolar and Matripolar Primate Lineages

Africa appears to be the site of this dual evolution of primates to the higher apes, through pre human hominids to modern humans. The DNA obtained from various humans living on the continent of African is by far more variable in sequence than DNA taken from humans on any other continent in the world. This strongly supports the African origins of all humanity. There is a debate about whether apes migrated out of Africa first to evolve into pre-hominids at other global locations, or whether complete evolution to the level of hominids occurred in Africa before any outward migration occurred. Or, whether both occurred and the modern forms, originating either within or outside of Africa, outcompeted previous forms to extinction, or mated with them.

Be Fruitful, Multiply and Replenish the Earth

The previously ocean-surrounded African subcontinent finally drifted into its present position and docked against Europe only a few million years ago. Before that event, primate migration out of Africa was impossible.

Figure 9 diagrams the post-ape, pre-human, and hominoid migrations into the eastern hemisphere. In that scheme, it may be seen that hominids migrated out of Africa to Europe and Asia several times, only to became extinct, the latest being Peking Man and Neanderthal types. Finally, in that doubtful version, a single more advanced Negroid race emerged and migrated out of Africa relatively recently and killed off all earlier competitors. We are all supposedly the descendants of these genocidal ancestors.

Observation #15: Patripolar and Matripolar Lineages Have Made It Into The Present

The idea that one hominid lineage took over the earth can only be half-true because, as indicated, not just one human lineage took over the world. In fact recent MRI work on the hemisity of 120 Caucasian subjects shows, not only that Familial Polarity opposites attract, but also that RPs have larger corpus colossal sizes than their LP partners in both familial polarities. These corpus callosum data strongly contradict the once popular belief that all humans derive from a single genocidal African stock. Modern society has yet to confront the possibility that two different human species actually live among us today.

Figure 9, Hominid Evolution: "Out of Africa" vs. the "Multiregional" Hypotheses

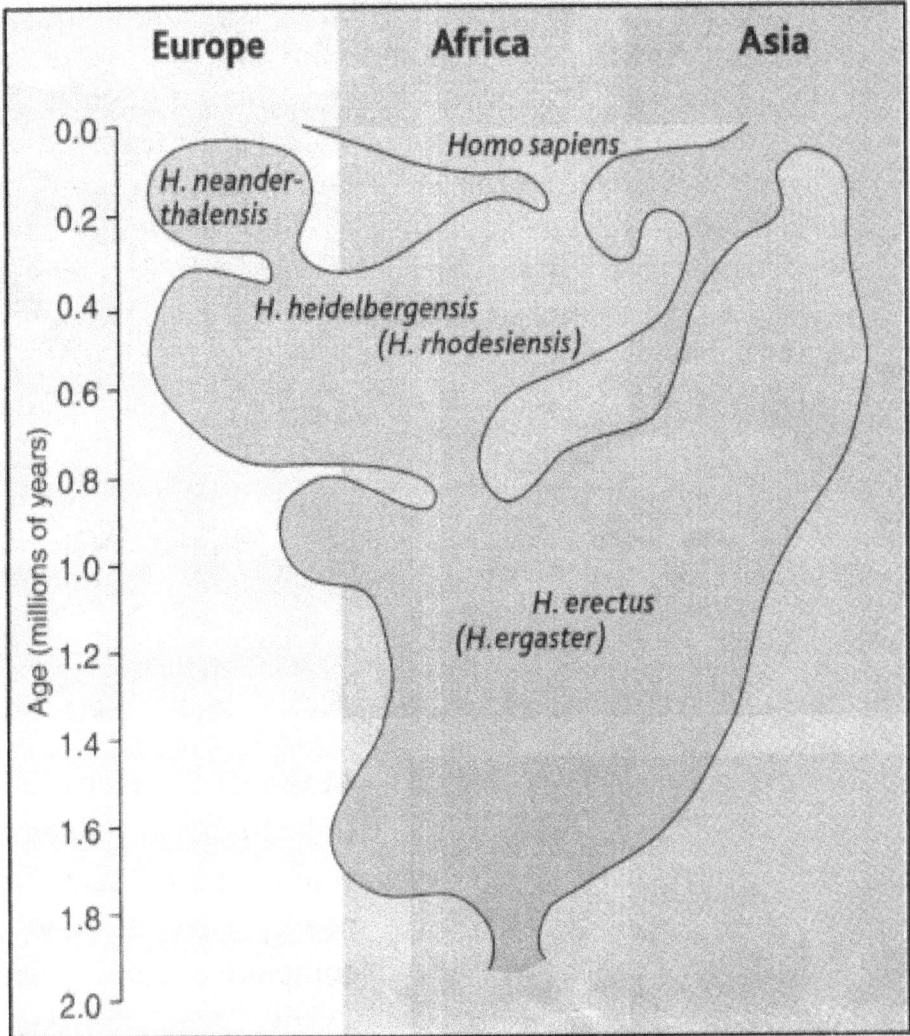

Go with the flow. One view of how various human species might have dispersed in space and time.

CHAPTER 2

That two pre-racial lineages emerged into the present to produce two contrasting human species populations creates a paradigm shift in our understanding of the nature of human relations, both in the past and especially today. This has yet to occur in today's world population. The rest of the book explores the effect of cross-breeding between matripolar and patripolar individuals to produce several types of human conditions, which at present have no scientific explanation for their origin, their genetics or prevention. These are dyslexia, homosexuality, pedophilia, and schizophrenia.

CHAPTER 3. Evidence Part 3: EFFECTS OF CROSS BREEDING

Observation # 16: Crossbreeding Produces Hybrids With Unnatural Neural Cross-wirings Causing Several Behavioral Abnormalities, or Maladies.

Currently we consider ourselves as one multiracial species. However, the discovery of Familial Polarity with its two opposite pre-racial true breeding lineages suggests otherwise. Consider the following additional evidence. When the childlessness among the four possible familial polarity couples was determined, the following result was found (84):

```
of the 40% who were RF-LM pairs,     15% were childless
of the 27% who were RM-LF pairs,      9% were childless
of the 20% who were LM-LF pairs,    *38% were childless#
of the 13% who were RM-RF pairs,    15% were childless#
* = significantly different.    # = birth anomalies in children.
```

What does the Spousal Hemisity Study Tell Us?

Based upon hundreds of observations, we know that about two-thirds of the families studied had spouses of the complementary hemisity (84). This was independent of their race. Either they were RM-LF families, or they were LM-RF families. Children of these family types usually had the same hemisity as the same gender parent. The one-third of the families with same hemisity spouses did not breed true and produced offspring, of either random (RM-RFs) hemisity or mostly LPs (LM-LFs).

Further, when the numbers of childless pairs were determined, only an average of 13% of the RM-LF, LM-RF, or RM-RF pairs were childless. A significantly greater 38% of LM-LF pairs were childless. Apparently, the LM-LF cross breed pair was sufficiently incompatible genetically as to elevate the childlessness rate three-fold for that pair. Spontaneous abortion due to genetic defects may be one of the reasons.

In addition, essentially all of the children from the hybrid LL or RR pairs have developmental behavioral anomalies. As a result, they are not like the matri- or patripolar wild types of their parents, as we shall see. If there were not cross-breedings between two different species, such hybrids should not be produced. This is an important point.

Many natural and artificial hybrids between animal species are known, each having characteristic reproductive defects due to chromosomal mismatches or other reasons. For example, these include the Jack donkey (62 chromosomes) and the horse mare (64 chromosomes) to produce mules (63 chromosomes). In these hybrids, the males are absolutely sterile, and females have borne young in only 60 recorded incidences since the year 1527. The mule is thought to be more intelligent than either of its parents are, and in certain situations is hardier as well. I have seen trained mules jump off a high dive.

In the opposite situation, a stallion breeds a jenny donkey to produce a hinney. This hybrid is significantly different in appearances and properties from a mule. In other hybrids, for example between the Przewalski horse with 66 chromosomes, and the domestic horse with 64 chromosomes, the males of the hybrid offspring are able to impregnate the domestic horse, while the females remain sterile. These observations of hybrid fertility would appear to have some relevance to the human situation. Thus, the elevated childlessness of the LL couples appears to be further evidence that two human species exist.

The remainder of this book will document the miswiring differences in the familial polarity hybrids resulting from RM-RF cross breeding and those from LM-LF crosses.

After the Last Ice Age

Shortly after the last ice age, family relations were relatively simple compared with today. For example, among the small nomadic populations of male dominant (patripolar, Chapter 2) big game hunters living in caves near the glaciers, the males were large, muscular, and dominating. Their females were considerably smaller than their men were, but very supportive. That is, in addition to being childcare providers, they were skilled game processers who produced fur clothing and leather for sleds, and boats from animal hides. Their vanishing fat-combusting, "flame technology" enabled them to live independent of wood for fire and heat. This technology was the process of the controlled burning of animal fat (blubber), which has a higher caloric density than wood, to heat their lodgings and to cook. The children were usually the same: robust, aggressive boys, and diminutive clever girls. The best hunters had their pick from among the similar young women who competed to serve him. For eons, they had endlessly produced replicas of themselves: dominant males and supportive females.

An example of this type of patripolar life has been preserved in the 1922 documentary film by Robert Flaherty (38) about Canadian Inuits, called: *Nanook of the North: A Story Of Life and Love In the Actual Arctic*. In 1989, this film was among the top 25 films selected for preservation in the United States by the Library of Congress as being "culturally, historically, or aesthetically significant."

In contrast, elsewhere after the last ice age there were also different small groups of survivors who were female dominant (matripolar, Chapter 2), and completely isolated from the male dominant stock. The family relations of these people were also a simple "boy meets girl" anciently to reproduce more of the same. However, these groups used the equally effective but opposite reproductive strategy of the matripolars. Here the female was the biologically dominant reproductive partner, with the male serving as the supporting high tech partner. In this case the males did not fight each other for leadership, as did the aggressive patripolars. But, instead they competed for the best territory. Then, the females selected from among the courting males holding the best territories. He was often the one who would climb the highest mountain, or swim the deepest sea for her ladyship. Thus, powerful females gained a protector and his assets.

As long as each of these two very different true breeding post ice age human subspecies were separate from each other, each anciently having migrated independently out of Africa much earlier, family matters stayed quite simple. But as the two populations thrived and expanded, population interfaces between them formed, because of their opposite biological and cultural differences. At first, they competed with each other. Patripolar males fought for their Fatherland, against matripolar males fighting for their Motherland, as some still do today.

However, as depicted in Jean Auel's anthropology-based book series, *The Clan of the Cave Bear"* (2b), migrations occurred that bridged the Matri and Patripolar population interfaces. Then, occasional cross breeding occurred. By the production of hybrid offspring, this critical event forever brought conflict and suffering into their lives. Further, the hybrids grew up and married other hybrids, producing a bewildering array of alternative mating options each with different anomalies.

Observation #17: It's No Longer Boy meets Girl, But 16 Kinds of Boys Meet 16 Kinds of Girls!

Now it was no longer a boy meets a girl from the same tribe. But boy or girl meets others from well over 100 mating options. That is, first from the overlap of the two species, came two additional types of boys and girls to mate with. However, those subsequent cross-species pairs produced hybrids, and the next generation these hybrids were added to the pot of mating possibilities. These formed a wide range of new possible interspecies combinations resulting in corresponding hybrids, the transvestite pedophile, or the schizophrenic being the extreme. Since in areas of matripolar-patripolar overlap cross breeding was occurring in major cities at least as early as the Greeks, the spectacularly gifted wild types now are in the minority, with most of our associates being hybrids of some kind.

With our million year BC genetic lag, we are not well equipped for such mating choices. Many of them result in incredible marital conflict upon many levels, from the biological to the cultural. Another Inuit movie, *Atanarjuat: The Fast Runner*, a 2001 Canadian film (72), produced entirely by Inuits, elegantly illustrates this. Set in the ancient past, the film retells an Inuit legend of (unrecognized patri-matripolar) conflict down through centuries of oral tradition. It is included in a list of Canada's Top Ten Films of All Time.

Furthermore, because these patri-matripolar cross-breeding hybrids are unrelated to race, it is no wonder that recognition of "Familial Polarity" has taken so long. Racial differences are far smaller than familial polarity species differences. That is, all races contain both elements. Consider the recent genocide of patripolar Tutsi Africans by the matripolar Hutus. Until now, we didn't have a clue about what was going on.

Thus far, it is becoming clear that at least five different general hybrid types result from the patri-matri cross breeding and they are illustrated on the front cover of this book. The dyslexics, the trans-heterosexuals, are the product of RM-RF crosses, while the products of the cross-breeding LM-LF couples include, cis-homosexuals, and trans-homosexuals. Then, cross-breeding of second or higher generation hybrids produces even more miswired offspring including pedophiles and schizophrenics.

It appears that at this point in human history there are more hybrids than pure matripolar or patripolar stock. Thus, existing pure species individuals often rise to the top of human endeavor becoming admired leaders in many fields, often being multilingual and multitalented. This appears to be due to their unmatched memories.

The Neurobiology Underlying the Familial Polarity Hypotheses of Cross-Wired Hybrids

The nervous system of all bilateral animals, including humans, is composed of two populations of neurons: those with targets on the same side of the body and those with targets on the opposite side. For all neurons, the decision of whether to send their axonal fibers across the midline is universal. These characteristics make it an extremely interesting biological model for developmental neurobiologists, so much so, that it is among the best studied topics in brain development.

At the midline of the developing human spinal cord during pregnancy is a transient structure composed of ependymal cells, called the floor plate. The floor plate plays a critical role in patterning neuronal fates and projections within the spinal cord. A similar floor plate for the brain is to be found in the transcerebral crossing called the corpus callosum. These plates are the source of factors that specify neuronal cell fates as well as guidance molecules that orient the growth of their axons during pregnancy.

What are the signals at the midline, and what are the receptors on the axonal growth cone that control whether axons should cross or should not cross? After crossing once, what mechanism prevents the growth cones from crossing again? Studies have led to the suggestion that the midline secretes repellent, as well as attractant, stimuli and that the decision of crossing is regulated by the way growth cones interpret and balance this symphony of diverse stimuli.

Axons from crossover neurons are initially drawn to the midline by attractant proteins, which include members of the netrin family. However, after crossing, these growth cones lose responsiveness to netrins and become sensitive to repellents made by midline cells, which include Slit proteins. Slit is a repellant for non-crossing neurons or for once-crossed neurons to prevent re-crossing. In mutants, many axons abnormally cross the central nervous system (CNS) midline, with some doing so multiple times.

To this complexity, add the unrecognized presence of Familial Polarity interbreeding with the potential for developmental path finding errors to occur in hybrids during pregnancy. These errors are proposed to be due to the absence of some of these needed midline factors or to their excessive presence. Such cross-wired hybrids are the possible consequences of two closely related species cross-breeding to produce hybrids, which then further cross-breed to produce a large array of possible combinations.

The Familial Polarity Hybrid Hypothesis asserts that each of these fetal hybrids lack required factors that are required in both the sperm and ova from one, or the other of the two species. Genetic differences for fetal development appear to exist between the two species, so that when cross-breeding occurs, there are developmental structural anomalies, resulting in a deviation from wild type wiring and behavior. Some of these cause little problems, while others produce serious aberrations. The range of these developmental anomalies, addressed in this book, is illustrated symbolically in **Figure 10**.

Personal Suffering is Brought By Ignorance of Familial Polarity

In part due to the perceived futility and irreversibility of their cross-wired behavioral tendencies, there is a higher mortality from suicide, drug abuse, and AIDs among the trans-species hybrids. Many of the lesbian, gay, bisexual, and transsexual (LGBT) community, have suffered enormously. Although unaware of their own origins, these hybrid groups are beginning to aggregate into politically significant groups influencing global culture. Through the permissiveness of political correctness, large blocks of LGBTs have become active in entertainment media and in government. In contrast, hopefully more pedophiles are being weeded from our clergy and coaches.

Enormous revulsion and prejudice among the general population to the horrific and deathly cultural orientations of the goths, homosexuals, and pedophiles leads to their visceral rejection by normals. Partly in revenge, some in the LGBT community intentionally attempt to shock them. *The Rocky Horror Picture Show*, featuring trans-sexuality and cannibalism was ignored at

Figure 10

Hybrid Familial Polarity Midline Crossing Pathologies

| M F | M F | M F | M F | M F | R L | R L | S T | S T | U V |

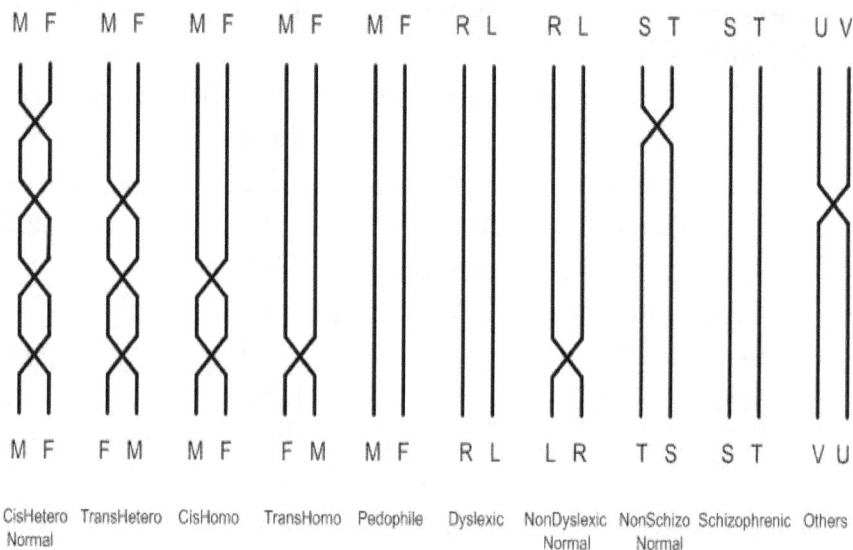

| M F | F M | M F | F M | M F | R L | L R | T S | S T | V U |

| CisHetero Normal | TransHetero | CisHomo | TransHomo | Pedophile | Dyslexic | NonDyslexic Normal | NonSchizo Normal | Schizophrenic | Others |

first. Now, it has become the longest running cult film in history, thanks to members from the LGBT fraternity.

Hybrids have Egos too and are usually highly intelligent. Their ostracization in high school by rejecting non-hybrids has resulted in outbursts of murderous outrage and desire for revenge, as in the Colombine High School massacres and other similar incidents.

Recently, major musical performers who are transvestites, goths and/or pedophiles have attracted millions of loyal followers, especially among the formerly disenfranchised and downtrodden familial polarity hybrids, who through them, have discovered that they are not alone. The natural weirdness and audacity of these entertainers grabs the attention of everyone, especially kids and teenagers of all kinds. This has had a major impact upon the entertainment industry and thus upon society.

In Latin America and elsewhere, Gothic type gangs have arisen, whose members can be almost totally blue with facial, neck, and body tattoos. They thrive on terrifying the populace with the horror of dismemberment in the process of feeding their drug needs and practicing their Satanic beliefs. Thus, they extort the population with kidnappings, often planned behind bars by jailed leaders. Not uncommonly, these end in death of the victim. Initiation of admiring homeless urchins into the gang often requires them to kill as proof of commitment to their new family. Membership is forever. If a member ever tries to leave the gang, one of their friends will murder them. This is a very destructive parasitic subculture.

And who speaks for the self-terrorized schizophrenics with their delusions of reality and hallucinations that can lead to violence.

The unrecognized hybridization between the two human species is the topic of this book. The goal is to bring familial polarity into awareness. That way we can begin to think about it. This will enable us to minimize its serious damages and optimize its significant benefits. As will be seen, hybrids have been among the great contributors to human culture.

CHAPTER 3

CHAPTER 4. WE DYSLEXICS ARE CROSS-WIRED HYBRIDS

The word, dyslexia, literally means poor reading. The term developmental dyslexia (dyslexia) refers to poor reading present *in a person of normal or better intelligence who has no medical, psychological, or socioeconomic condition to account for their difficulty in reading.* Thus, dyslexia is a broad term defining a hereditary learning disability that impairs a person's fluency and comprehension in being able to read and write. It shows itself as a difficulty in connecting the symbols of language with their spoken sounds (phonemes, i.e., the subject of phonics). Underlying the dyslexic deficits is a slowed processing rate, an impaired short-term memory, and a spotty, selective long-term memory that leads to spelling and other habit forming difficulties. Dyslexia is not a developmental lag and does not resolve over time.

Dyslexia was first recognized in the late 1800s as "Congenital Word Blindness." In the early 1900s, dyslexia was proposed to be the failure of the brain to establish left hemispheric language dominance. In the late nineteen hundreds, dyslexia was suggested to result from abnormal cortical development during *fetal* brain development. The corpus callosum has also been implicated. More recently, certain brain activity differences have been detected by functional MRI, as will be discussed later. However, no specific mechanism for the origin of dyslexia has existed until the one proposed by the author of this book.

It has been estimated that between 5 to 30% of the population are dyslexic. The Connecticut Longitudinal Study, representative of all children attending public kindergarten in Connecticut, assessed reading and intelligence in each child in the study and found that one in five children was dyslexic (110c). Since only 5% of children receiving special education services are classified as dyslexic, a huge number of struggling readers are not identified in their schools.

Amazingly, dyslexia is not an intellectual disability since it and IQ are unrelated. In fact, for some who by hard work have overcome their dyslexic disadvantage, it has turned into "the dyslexic advantage". Thus, some adult dyslexic individuals, such as Albert Einstein, Charles Darwin, and others to be noted later, have been brilliant "outside the box" big picture thinkers. In fact, entrepreneurs, including billionaires, are five times more likely to suffer from dyslexia than the normal population. In marked contrast, there is essentially no dyslexia among middle managers, compared with the general population (32).

Dyslexia is often first noticed in a child as a delay in speech or in reading. Intense extra phonics training greatly assists the child to learn to read and write. This avoids later adult failures in literacy. Einstein didn't speak until he was six. My own commercial pilot son did not read until he was eight. Imagine how he must have felt going to school for five years with classmates who could read by the age of three.

Signs and Symptoms: (derived from "Dyslexia-Wikipedia")

Preschool children

Delays in speech

Slow learning of new words

Difficulty in rhyming words

Low letter knowledge

Letter reversal or mirror writing

Early primary school children

Difficulty learning the alphabet, or letter order

Difficulty associating sounds with their letters

Difficulty in rhyming or distinguishing syllables

Difficulty sounding out words

Difficulty retrieving words and naming problems

Difficulty learning to decode written words

Difficulty distinguishing between similar sounding words

Older primary school children

Slow or inaccurate reading

Very poor spelling and ugly handwriting

Difficulty reading out loud, but may read well silently

Difficulty associating words with their meanings

Difficulty in time keeping and concept of time in doing tasks

Difficulty with organization skills (working memory)

Difficulty in sounding out the pronunciation of new words

Tendency to omit or add letters/words when writing or reading

Secondary School Children and Adults

The unique tedium required for learning the "three Rs" is always painful for dyslexic children. Some people with dyslexia are able to disguise their weakness (even from themselves, as I did!) and learn to read and write, erroneously thinking that everybody had to work as hard as they did to learn. Once the writing barrier is surmounted, they often do acceptably well in high school, and better than average at the college level. Some never encounter that threshold at which they are no longer able to compensate for their learning weakness.

A common misconception is that all dyslexic writers reverse letters or write them backward. This only occurs in a small population, although reversed letters in words typed from a keyboard is quite common.

Some Associated Deficiencies

Dysgraphia-effects hand eye coordination, such as tying knots, or carrying out repetitive tasks, such as letter writing automaticity. Habits are hard to learn and easily disrupted. This leads to clumsiness and require continued planning of acts that should have become automatic long ago.

Dyscalculia-Often people with this condition can understand very complex mathematical concepts but have difficulty retrieving basic math facts involving addition and subtraction.

Cluttering-Stuttering-Hesitant Speech A speech fluency problem involving both the rate and rhythm of speech, causing slowed speech or impaired intelligibility.

<u>ADHD</u>-Attention Deficit Hyperactivity Disorder is a separate malady from dyslexia. However about 10% of ADHD victims might also be dyslexic, as expected from the abundance of dyslexia.

Table 15 lists some of the many of the problems that a typical dyslexic will have to confront and over-come, either by rote memorization or by compensated avoidance in order to become a functioning member of society.

Dyslexics Have Specific Memory System Problems

Without memory, the sound of a snapping tree branch above us occurs for a second and then is gone for-ever. Without memory of this event we cannot respond to it and are utterly helpless to save ourselves from the limb crashing down upon us. **Figure 11** is a diagram of our present understanding of the human memory systems. Memory is divided into long-term memory and short-term memory.

Types of Short-Term Memory: Animal and Human

Short-term memory keeps the sound of the falling branch above us in a repeating, reverberating form lasting a few seconds, during which we can focus to decide how to get out of the way. There appears to be two different short-term memories. The first is the low capacity and briefly lasting, presently unrecognized subcortical short-term memories from our animal past, presently called short-term memory, but more appropriately named sub-cortical short-term memory.

The second, called working memory (102), is a larger, longer lasting cortical form of short-term

Table 15. Typical Difficulties that Dyslexics Often Face:

1. Delayed learning to read and write. Difficulty connecting words and syllables to letters.
2. Once they learn to read, difficulty in reading up to speed, especially when reading aloud.
3. Very slow. Lack of automation requires continual planning and effort.
4. Problems remembering: Have a very short working memory. Must look at the number twice to dial regular phone numbers.
5. Difficulties learning proper spelling: impaired encoding, i.e., the writing of the word "close" for the word "clothes".
6. Handwriting remains poor, unsightly, or illegible.
7. Losing track of the location of personal items: credit cards, keys, glasses, and cell phones.
8. Difficulty recalling names and numbers: have unreliable auditory and visual memory
9. Major problems recalling oral directions or instructions. Must write them down
10. Calculations are slowed. But, they are accurate when extra time is allowed for tests. Difficulty in remembering math symbol definitions.
11. Slow of speech: Delayed starts. Rarely a glib speaker. May stutter.
12. Difficulty concentrating in noisy environments. Need silence to think.
13. Retarded rate in forming habits of all kinds. Habit learning is very slow for things from spelling to handwriting to body movements.
14. Dancing is not natural and very difficult to learn in adulthood. Their body appears disconnected from the music, perhaps like their letters once were to sounds. However, they enjoy rhythmic music greatly.
15. Difficulty in counting musical time, dissociated from feeling the rhythm.
16. Can experience movement reversals: push vs. pull a door open, reversing letters while typing.
17. Mental Blocks, or "choking" caused by fear, often psychosocial fear.
18. Disappointing high stress performance. They usually do much better during practice.
19. Accident prone under stress, resulting in a frozen or slow, inaccurate response.
20. Somewhat clumsy.
21. Difficulty in imitating others, need procedures to be broken down.
22. Extreme difficulty in learning a second language.
23. Can have a poor self image of never being good enough.

Figure 11. Human Memory and Its Deficits in Dyslexia

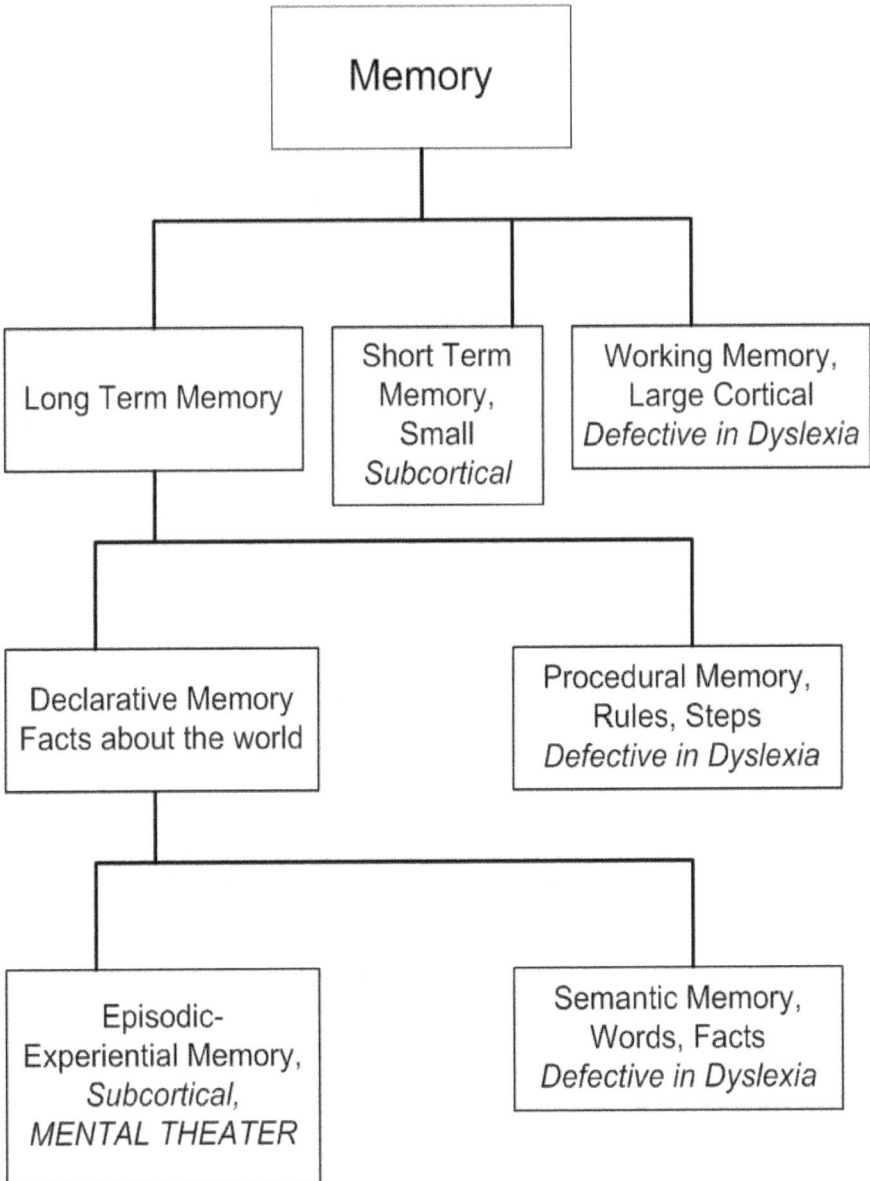

memory. It exists in humans who are not dyslexic. This is the higher capacity and more prolonged cortical memory normally required for the capability of language.

Unfortunately for dyslexics, cortical working memory is disconnected and has become unavailable. This leaves dyslexics only with subcortical short-term memory for their entire lives. Subcortical short-term memory appears to be located in the striatum and is hardly large enough to hold part of a telephone number. It also disappears quickly. If the dyslexic is to develop into a functioning human being he or she must find ways to compensate for this deficiency in working memory. Millions of people have accomplished this.

The Long Term Memory of Procedures

This brings us to long-term memory and its types. Amazingly, a long-term memory multisensory record is captured permanently from the brief, one-time exposure to what is held in short-term and/or working memory. The molecular basis for this incredible feat is, as yet, unknown.

One category of long-term memory is called procedural memory. This is the memory of habit formation and is to be seen at the heart of language, reading, and writing, as well as automatic actions in our daily lives. As you may have guessed, the formation of procedural memory habits is enormously impaired in dyslexics. This defect could be said to be at the root of most of their problems.

Anciently, the learning of a procedural routine was needed by a newborn cub to learn the way back to its den. In language it is required to form automatic connections between sounds and symbols. At a higher level, it connects meanings to words, and even higher, to the spelling of words (54). It is also used in the formation of habits, so important in moving from concentration to automaticity. Its absence makes dancing a non-automatic, effortful, and clumsy.

Declarative Memory: The Long-Term Memories of Experiences (Episodic Memory) and of the Conclusions Derived From Them (Semantic Memory):

A second form of long-term memory collects facts about our world and is called declarative memory. As might be expected from evolution, it has two forms. One, called semantic memory, is packaged cortically together with working memory and procedural memory. It is the record of the world of facts and figures abstracted from experience and devoid of context. The semantic memory system is greatly impaired in dyslexics.

A much older subcortical form of declarative long-term memory is called episodic memory. It is so called because it contains a complete record of episodes of personal experience, still held within their multisensory context. It is large and unwieldy and has been said to be primary memory, and it is proposed to exist within the cerebellum (93). This is the long-term memory that dyslexics can learn to access. It is in a story-like format: i.e., who did what to whom, when, why, where and how. Unlike the declarative memory derived from it, the story has not been abstracted nor its context removed. Given adequate processing time, it is a treasure lode of multimodal information. Thus, in conclusion, the dyslexic only has subcortical short-term and subcortical long-term memory available from which to survive.

Neuroanatomical Observations of Differences in Dyslexia

Past thinking regarding the mechanism for dyslexia contained disconnected facts, most of which were true. In addition to suspicions about lack of LH language dominance, there was the discovery that Einstein's brain was different from a normal person's brain in the same way as a dyslexic's brain is. That is, both are more symmetrical, with a smaller left language cortex (planum temporal) and larger adjacent spatial cortex (parietal). Einstein complained about his poor memory for words, his inability to learn foreign languages, trouble remembering math facts, and keeping the beat when playing his violin. Sound familiar? In contrast, like other dyslexics, he had an unusual ability to visualize complex spatio-temporal relationships, thought to be processed in his 15% enlarged parietal lobe.

More recently, the use of functional Magnetic Resonance Imagery (fMRI) has revealed that the dyslexic brain is working less intensely than normal in the more rearward planum temporale language region. But, it was over activated in the frontal cortex, working over four times as hard than normal. It is as if the front is trying to compensate for a deficit in the rear. Thus, brain imaging gives visible evidence of the reality of dyslexia. Proper training for reading partially reversed this imbalance.

However, in the absence of an insightful context, turning these observations into a compelling theory has been very difficult. As one might imagine, the discovery of the existence two closely related human species, differing primarily in their reproductive strategies, provides a new realm of possibility. In this chapter, the first specific mechanism for the cause of dyslexia is presented.

Observation #18: All Right Brained Children From R-R Crossbreeding Parents Were Dyslexic

This was observed hundreds of times (84). Their left-brained siblings were not.

Observation #19: All Dyslexics Are Right Brain-Oriented RxPs

When the hemisities of the offspring from the four possible hemisity pairs were surveyed and tabulated, a surprising result emerged: All of the RxP hybrid offspring between patripolar RM and matripolar RF partners were dyslexic (84)! In contrast, the hybrid LxPs offspring of the RM-RF couple were not dyslexic. In fact, no LP has ever been found to be dyslexic. LxPs will be addressed in the next chapter. In our nomenclature, the insertion of an w, x, y, or z between capitalized normals refers to a specific type of familial polarity hybrid. Here x refers to children of R-R parents.

Thus, when the sperm from a RM fertilizes an RF ovum, four outcomes are possible. They are that the child will be a non-dyslexic LxM or LxF. Or, the child will be a dyslexic RxM or RxF. Homozygotes are required for both wild types. Heterozygotes fail due to excess or lack.Since in our population (see Chapter 4) 13% were RM-RF couples, this suggests that their RxP offspring should account for roughly 7% of the population. As mentioned, the abundance of dyslexics in the population had been estimated previously to be between 5 and 20%.

The Familial Polarity Hybrid Theory of Dyslexia:
     As illustrated in the diagrams of a LP, a RP, and a RxP in **Figure 12**, the RxP is a familial polarity hybrid who appears to have his or her memory bundle "screen" cross-wired to the left side. The memory screen appears to be composed of working memory, procedural memory, and semantic memory. These cortical memory systems are no longer accessible to the RH executive. As a result of this loss, it must depend for memory solely upon its remaining subcortical sources (short-term memory and episodic-experiential memory).
     What might have happened during a cross-polar pregnancy to cause dyslexia, and how does that produce dyslexic behavior? First, letter reversal and hemispheric memory laterality evidence in **Figure 12** suggest that a cross-wiring defect must have occurred. This would be expected, due to the presence within the fetus of two different and somehow non-matching chromosome sets, one patripolar contributed by the RM father, and one matripolar from the RF mother. This book is about the multiple possible cross-wiring consequences in utero of hybrids, dyslexia being only the first. The author had earlier proposed that the location of one of these important failed central nervous system crossovers occurrd in the cerebel-

Figure 12:

ORIGIN AND MECHANISM OF DYSLEXIA

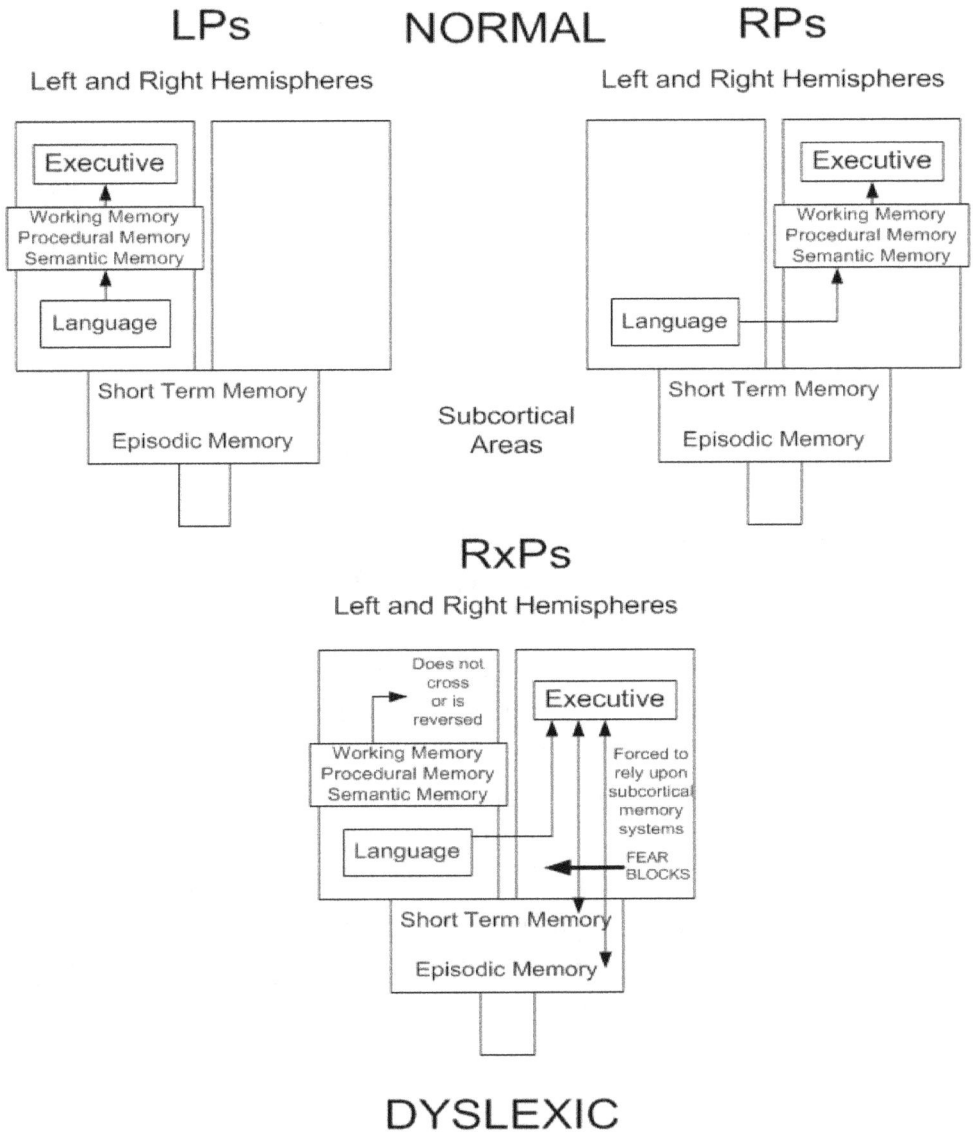

LPs   NORMAL   RPs

Left and Right Hemispheres   Left and Right Hemispheres

Executive

Working Memory
Procedural Memory
Semantic Memory

Language

Short Term Memory

Episodic Memory

Subcortical Areas

Executive

Working Memory
Procedural Memory
Semantic Memory

Language

Short Term Memory

Episodic Memory

RxPs

Left and Right Hemispheres

Does not cross or is reversed

Executive

Working Memory
Procedural Memory
Semantic Memory

Language

Forced to rely upon subcortical memory systems

FEAR BLOCKS

Short Term Memory

Episodic Memory

DYSLEXIC

lum at the superior cerebellar peduncular decussation (Morton, 2003a).

Figures 12 and 13 further illustrate the basis for difference between normals with photographic memories and dyslexics with variable performance memories. As may be seen, a purebred RP normally has a photographic memory because his or her executive is on the right immediately next to his or her memory module "screen", which is directly connected to three primary cortical memory banks. Wild type RPs do not have to memorize names or phone numbers because they can capture names and numbers directly off the adjacent memory screen on the right side of their brain, even if they have only heard them once. To us, their normalcy is giftedness.

Observation #20: In Dyslexics, the Cortical Memory Module is Displaced to the Left Hemisphere.

It is incredible that in familial polarity hybridization cross-wired dyslexics, all three of the cortical memory systems migrate inappropriately to the LH, as if they were a modular package. This leaves the right brain executive separated from access to the three cortical memory elements. One might think that the memory dislocated to the left should cross the corpus callosum through some new pathways. However, because its fibers of crossing only represent 1% of all cortical neurons (1), this appears not to occur extensively. However, that it does occur with letters and numbers, as illustrated in **Figure 13,** argues that cortical memory access is indeed cross-wired to the left.

Figure 13
Dyslexics Are Cross Wired Familial Polarity Hybrids

RP Normals: Executive Has Full Memory Access on the Right

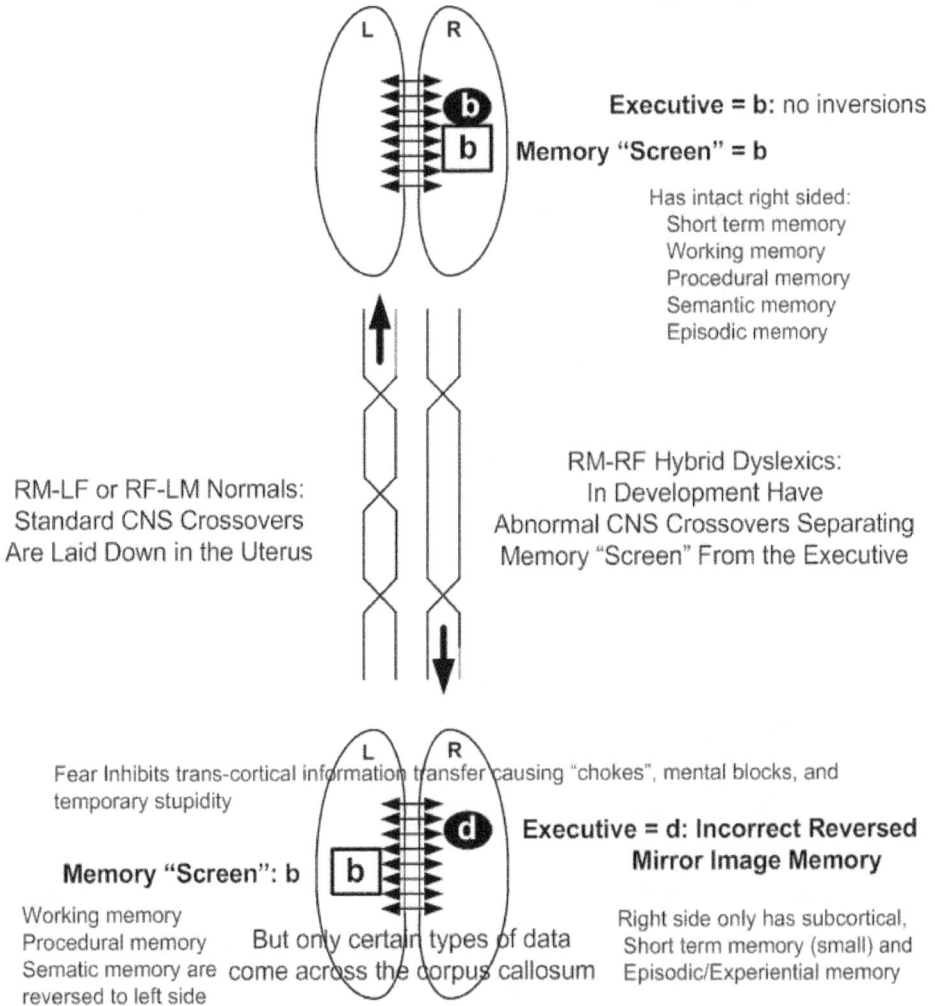

Executive = b: no inversions

Memory "Screen" = b

Has intact right sided:
Short term memory
Working memory
Procedural memory
Semantic memory
Episodic memory

RM-LF or RF-LM Normals:
Standard CNS Crossovers
Are Laid Down in the Uterus

RM-RF Hybrid Dyslexics:
In Development Have
Abnormal CNS Crossovers Separating
Memory "Screen" From the Executive

Fear Inhibits trans-cortical information transfer causing "chokes", mental blocks, and temporary stupidity

Memory "Screen": b

**Executive = d: Incorrect Reversed
Mirror Image Memory**

Working memory
Procedural memory   But only certain types of data
Sematic memory are come across the corpus callosum
reversed to left side

Right side only has subcortical,
Short term memory (small) and
Episodic/Experiential memory

RxP Dyslexics: Only Have Subcortical Memory Systems

Thus, for the right sided executive, only the sub-cortical resources remain; that is, short-term memory whose capacity normally is insufficient for language, or procedures, and the vast and unwieldy episodic memory from which to attempt to retrieve facts about the world. Yet, this situation actually confers a dyslexic advantage to those who have mastered reality through these two memory windows. They have become among the most creative and productive people in the world.

The Familial Polarity Cross-Wiring Hypothesis for Dyslexia: How It Accounts For Dyslexic Behavior

To have access to the facts, which are directly accessible to RP normals with their memory on the right, requires that the dyslexic laboriously memorize these facts by rote rehearsal. Here it is proposed that memorization may represent the effortful construction of a new artificial memory bank in the RH near the executive. Building and using this homemade memory system to substitute for the displaced original memory system, requires energy and time and may account for the much higher frontal brain activity of the dyslexic brain having resorted to such a system.

On the other hand, the higher frontal activity in the dyslexic brain may only be due to their lack of procedural memory access forcing them to concentrate hard on what normally is automatic and easy. In dyslexics, procedural memory is very impaired, requiring many more than ten times the number of repetitions than normal to form a habit. Thus, instead operating mostly on automatic habit behavior, and thus free to think ahead, the poor dyslexic has to overload their small short-term memory to concentrate on what he or she is trying to do. Regardless, the dyslexic must memorize endlessly to form a habit or to learn something at any level. However, the resources they must develop to memorize categories, also gives them a deeper understanding of the nature of order and of organization. Often, this brings insights not common to normal persons.

"The Dyslexic Advantage"

A book of the above name (32) is filled with examples of successful dyslexics. The authors postulate that by overcoming their disabilities, dyslexics often accomplish much more than talented normal people do who have never have to face such difficult challenges. Of course these overcompensation-based accomplishments must be sorted from the global skills conferred upon all dyslexics by naturally being right brainers. **Table 16** lists a few famous people with dyslexia.

Non-dyslexics excel at precision, accuracy, efficiency, speed, automaticity, reliability, replicability, focus, conciseness, detailed expertise, applying rules, and spotting differences. Generally, these are the skills dyslexics struggle hardest to develop.

In marked contrast, dyslexics see the gist or essence, the larger context behind an idea, with a multiple dimensionality of perspective. Because they see commonalities, they see new unusual or distant connections, use inferential reasoning, detect ambiguity, and can recombine things in novel ways. They have a generally

Table 16. Famous People With the Gift of Dyslexia:

Dyslexic Inventors and Scientists and Physicians:

Thomas Edison

Alexander Graham Bell

Charles Darwin

Albert Einstein, Ph.D.

Michael Faraday, Ph.D.

Nils Bohr, Ph.D.

Harvey Cushing, M.D.

Jack Barchas, M.D.

Dyslexic Political Leaders:

George Washington

Thomas Jefferson

Winston Churchill

John F. Kennedy

George H. W. Bush

George W. Bush

Dyslexic Entrepreneurs and Business Leaders:

Henry Ford

Charles Schwab

Ted Turner

Dyslexic Athletes:

Babe Ruth

Muhammad Ali

Bruce /Caitlyn Jenner

Magic Johnson

Dyslexic Artists:

Leondardo da Vinci

Pablo Picasso

August Rodin

Dyslexic Filmakers:

Walt Disney

Steven Spielberg

Dyslexic Actors and Entertainers:

John Lennon

Tom Cruise

Harrison Ford

Whoopi Goldberg

Anthony Hopkins

Jay Leno

Suzanne Sommers

Robin Williams

Salma Hyek

inventive mindfulness and attention, resulting in creativity, insight, and wisdom.

Dyslexics can adapt whatever materials they find at hand to construct their projects, demonstrating an unusual ability to see analogies between the needed unit and "spare parts available in their surroundings". They have strength in shifting perspectives and in global thinking. They can combine different types of multidisciplinary information to create a parsimonious, integrated, and unified whole.

Some dyslexics have the ability to predict past or future states accurately, especially when components are unknown or ambiguous. They use intuition and insight based upon general knowledge and past events to produce sudden, fully formed answers. Later, they must then determine the route of their logic by working backward to fit the data.

Ways to Assist Dyslexics in Compensating for Their Defect

The dyslexic needs to be identified as young as possible, and somehow informed that, through no fault of their own they were born with a memory defect, compared with others with photographic memories. They need to be reassured that the time-consuming repetitive process of memorization of facts and the repeated practice of procedures can overcome this defect.

Early assistance by extra reading training is essential to help them to become able to read at a useful rate. According to the National Reading Panel (94b), this enrichment should include five elements: 1. Phonemic awareness: the ability to be aware of, notice or manipulate the sounds of spoken language. 2. Phonics: learning to link letters to the sounds they represent. 3. Fluency: the ability to read both accurately and rapidly, and with good intonation. 4. Vocabulary: to understand the meaning of words read. 5: Comprehension: to understand and discern the meaning of connected text.

Further, these components are best taught via a systematic approach where instruction is intense (small groups of 4-5 students), with sufficient time (60-90 minutes/day). This supports them both to reach fluency for their grade level and gives them the opportunity to demonstrate their hard earned skills.

As children, they need to be encouraged that some of the greatest people in the world were blessed with dyslexia. And, that they too, to a great degree, can overcome their reading disabilities through hard work, memorization and practice, thus enabling them to enhance their lives and make the world a better place.

Non-dyslexics with photographic memories will have more time for an active early social life than hard working dyslexics will. If a dyslexic demands such an early social life, it will take away from the memorization time needed to succeed in their professional training. As a result, the dyslexic will lose out in competition for high paying jobs or for being selected as a mate by successful partner. Therefore, as the typical accomplished dyslexic has often said, "It was 10% intelligence and 90% horse power that enabled me to succeed."

Of course, later there will be plenty of time later for the dyslexic to have a rich social life. Often in intermediate or high school, "normals" are very popular because of their natural gifts. They tend to marry, have children early, find jobs, and drop out of the educational system before completing their professional training. Thus, the remaining dyslexics preferentially concentrate into the professions, and at times, out earn the normals.

Observation #21: Fear Causes Dyslexics to Choke with Mental Blocks.

Although currently unrecognized, dyslexics are those persons who suffer from the incapacitating mental blocks of so-called "Performance Anxiety" or "Choking". With the separation of the adjacent cortical memory screen from the right Executive Ego in these people, cerebellar episodic memory becomes crucial to performance. Fear inhibits the cerebellar cortex to produce "mental blocks" to the access of their information base. This leaves the dyslexic without data and thus temporarily stupid when they need their wits the most. This occurs because under immediate conditions of high anxiety, the locus coeruleus releases norepinephrine to inhibit the cerebral, cerebellar, and hippocampal cortices. This causes upset, decorticate behavior, because when the cerebellar informational support is blocked the reptilian brain stem instincts must take over.

Under situations of personal stress and fear, the dyslexic's compensatory data access system temporarily collapses, leaving the person cut off from their memory screens, in a "mental block". A few seconds after this breakdown, they often recover their thoughts, which may well have been brilliant. In contrast, while normal RPs,

LPs and certain other hybrids may experience debilitating performance anxiety, they do not go blank under pressure as RxPs do. Rather, their performance is improved. However, some are vulnerable to Type A stress disorders.

Mental blocks can decorticate dyslexics in any condition of high anxiety. These include during speeches, piano recitals, intimidating conversations or oral exams, timed test taking, golfing, motorcycle riding, and many others. Surprise and panic caused the author, who is an excellent superbike rider, to freeze and crash on several occasions. He did so after making thousands of perfect high-speed turns in familiar territory, before and since those incidents. This "blanking out" phenomenon appears may also be related to the norepinephrine based reductions of the vital frontal lobe activity of dyslexics. It has dramatically been visualized by SPECT (single proton emission computerized tomography) in patients showing performance decrements under pressure (2).

For a more typical example, as an academic I was once involved in a union meeting where our faculty annual contract was being negotiated. I had a relevant idea to share. When my hand was finally recognized and I stood to speak, I was suddenly face to face with a room full of attorneys and administrators. In my fright I "choked", went completely blank, and sat down in consternation. A few seconds later, my mind cleared and my idea was again clear. I leaped to my feet again and made my point.

Finally, regarding dyslexic dysgraphia: although I set a world distance record flying a hang glider whose control was accomplished by body weight shift, I never could master flying R/C model aircraft because of my poor finger-eye coordination (dyslexic dysgraphia). I would always crash them as soon as I built them. This may be related to the fact that my handwriting is incredibly messy with reversals and mark outs.

Conclusions:

Dyslexics are fully fertile familial polarity hybrids with unnatural cross-wiring of their memory system. This causes them to have to compensate greatly and work unusually hard to thrive within normal society. Some dyslexics have become geniuses who have greatly contributed to the optimization of human survival.

Statement of Respect for Hybrids and Their Rights:

Because of the two human species, hybrids exist. I am a hybrid dyslexic. It is not my fault. None of us can choose our parents or the time of our conception; neither the wild-type patripolars, wild-type matripolars, dyslexics, homosexuals, pedophiles, nor the schizophrenics. We are all human beings and deserve fundamental respect as such. Within the clarity provided by familial polarity, we now have a much more accurate understanding of our nature and their origins.

No disrespect to any hybrid group is intended in this book. However, in order to provide clarity to these studies, it must be stated that all hybrids including myself are mis-wired, compared to the wild types. This mis-wiring indeed can produce specific behavioral shifts away from the most efficient survival optimization of humanity. In terms of overall survival of the fittest humans, such mis-wirings appear to be abnormalities, defects, or deficits.

However, paradoxically, the hybrid's search for normality often results in super achievement that often accomplishes greater survival optimization to humankind than that contributed by complacent wild types. As will be seen, hybrids have become philosophers, geniuses, scientists, world leaders, and media stars who have changed the course of history.

Further, many of these abnormalities are of little immediate importance or are surmountable by proper application of information. They need not deprive any of us hybrids of meaningful, productive, and rewarding lives, including pedophiles and schizophrenics.

It is believed that by shifting away from previous unproductive points of view about the nature of dyslexia, homosexuality, pedophilia, and schizophrenia, the two species paradigm of familial polarity will advance our understanding. This will reduce the suffering of all hybrids and bring to wild types an understanding of the value that hybrids bring to humanity.

CHAPTER 5. TRANS-HETEROSEXUALS EXIST, UNRECOGNIZED

In the last chapter we discovered that the right brain-oriented familial polarity hybrid children (RxMs and RxFs) from the sperm of a patripolar man (RM) and the ovum of a matripolar woman (RF) end up dyslexic as a consequence of unnatural crossings during the laying down of the CNS in the womb. It was also found that none of their left brain-oriented children was dyslexic. So, the question arises: were the LxP offspring normal? The answer is no. To understand why, we must first discuss some issues related to sex and gender.

The Three Independent Components of an Individual's Sexuality

There are at least three independent aspects to one's sexuality. First and most obvious is the anatomical sex of one's *body*. In general, one's body is either male with a penis, or female with a womb. For millennia, wild type male and female (heterosexual) couples have endlessly reproduced almost identical copies of themselves by sperm fertilization of ova. This was true for all mammals, including both the patripolar and matripolar hominid lineages.

A second aspect of sexuality is the sexual identity of the object of one's reproductive *attraction*. The person with whom one becomes infatuated and to whom one desires reproductive closeness. In wild types, this is a same species person of the opposite body sex. As we will see in later chapters, hybrid cross-wiring can alter this instinctual choice to other unnatural non-reproductive targets. This type of alteration is the origin of homosexuality, where in general, gay men are sexually attracted to adult men, and where lesbian women are sexually attracted to adult women. Or, the alteration may result in pedophilia, where the unnatural object of sexual attraction has become a vulnerable child.

An independent third aspect of sexuality, sometimes called gender, has to do with one's own subconscious mental sexual *identity*. Does one identify oneself internally more as a man or more as a woman? Normally, men feel masculine and women feel feminine. However, with all the cross-wiring possibilities within familial polarity hybrids, things can become reversed, and reversed in more than one way. As we will see later, this results in the birth of quite a number of "non-normal" sexual configurations. This is not surprising if one recalls that most of the abnormalities shown by animal hybrids were in the area of reproductive fitness, i.e. highly able mules could not reproduce.

Observation #22: LxP Hybrids Are Trans-Heterosexuals With Reversed Sexual Identities

Now to the original question: all of the hybrid left male (LxMs) and females (LxFs) born to the RM-RF couples were heterosexual. The men were normally attracted to women and vice versa. However, what was different about the LxP hybrids was they all appeared to have reversed sexual identities. That is, LxM individuals with male bodies tended to identify with the softer side of life. In contrast, the LxFs tended to be more aggressive.

These were not the strident trans-sexuals, consciously asserting that they were born in the wrong bodies and in need of a sex change surgery. But, instead this reversal appeared to be expressed subconsciously, and because they retained a predominant heterosexual orientation, it was hardly noticed as more than variability of sexual expression.

Did you ever notice that among the population, besides stereotypic males and females, there are heterosexual men who seemed slightly feminine and heterosexual women that seemed somewhat masculine? These are the unrecognized slightly effeminate men and slightly aggressive women all around us. Here, these will be called trans-heterosexuals (TransHets). Trans is a term from chemical nomenclature where cis- refers to a structure the same side of an axis, and trans- refers to structure on the opposite side. In this nomenclature, normals are cis-heterosexuals (Cis-Hets) whose sexual-gender identity is the same as their body sex instead of opposite. In trans-heterosexuals, their sexual identity is opposite to the sex of their bodies. The relative abundance of LxPs should be about the same as their dyslexic siblings, that is 7% of the general population.

LxMs: Trans-Heterosexual Males (Effeminate men)

So, what is different about trans-heterosexual men from the normal male stereotype? In the US, these men (LxMs) are tend to be among the more artistic, sensitive, and intellectual heterosexual males. Hundreds were productive members on the faculty of my state university. These are the fathers, husbands, uncles, brothers, and sons that sometimes wear longer hair occasionally in a ponytail braid, and are generally non confrontational and mild while being non conformational. They love to cook and garden. Somehow delicate, it is uncommon for them to be attracted to hard manual labor.

In exaggeration, they have been called geeks, wimps, and mocked on television, as in the *Big Bang Theory*. One of them, Mister (Fred) Rogers, the beloved children's television host had a gentle, soft-spoken personality, and directness to his audiences.

As other example, in Star Trek, a standard RM, William Shatner, played Captain Kirk with a flair. In contrast, Dr. Spock, played by Leonard Nimoy, appears to be a stereotyped LxM. The contrast between Shatner and Nimoy goes beyond RM vs. LM differences. These LxM men can have an inherent female type of speech pattern and "body English" movements that identifies them to the observant. Johnny Depp is a current LxM movie star, well characterized by his more androgenous style, as in *Charlie and the Chocolate Factory* and *Alice in Wonderland*. Yet, he has been nominated several times as the sexiest man alive.

Some Women Choose More Effeminate Male Faces

The theory that women like feminine men is a fairly new one. Men and women alike presume that women like the rugged, athletic, and muscle-bound men. But a variety of unrelated studies supported the idea that women like feminine men. In one popular internet study done in the United Kingdom, men with traditionally feminine features were rated as being better for long-term relationships. Judging digitally altered photos, participants chose the more feminine-looking men as warmer and more committed (25b). Men with more masculine features were seen as better for short-term relationships.

However, there is more to these observations than meets the eye. Evolutionarily speaking, dominant matripolar RFs should be repulsed by the also dominant patripolar muscleman RM types of the opposite species, who could overpower and dethrone them. They are attracted to the more docile LMs. The reverse is true for LFs who want a masculine RM man to defend her. It is predicted that if the hemisity of the women were presorted for such an experiment, two groups of women would be separated, RFs attracted to and LFs repulsed by the softer males.

LxFs: Trans-Heterosexual Females (Masculinized Women)

In contrast, hybrid females LxFs from RM-RF couples tend to be noticeably more strident in their mannerisms. These are the mothers, aunties, sisters, wives, and daughters around us. In the US, they often are uncomfortable in culturally feminine attire. Rather, they tend to wear their hair shorter and wear shirts and slacks of coarser materials and use functional shoes, rather than the usual high-heeled style of females. Some of them tend to say, "It's a man's world" with envy, or "We can do it." as in the WWII poster. They often avoid cooking. They have an inherently male speech pattern and walk. Famous LxPs include U.S. Secretary of State Hillary

Clinton, Fox News Anchor Greta Van Susteren, Amilia Earhart, record setting pilot, and Angela Merkel, the Chancellor of Germany.

Some men enjoy such women because of their similar common interests. One such woman answered an internet advertisement from a man looking for a straight male-like female. She said that she understood the type of girl he was looking for: "It's hard to find others like me and I'm already married. I enjoy working on cars, fishing, working on a farm, any sport but basketball. I can shoot a gun and change my own oil. I have a hard time finding clothes 'cause they're always too "tight" (on top). I surprise people all the time. Either I surprise women by running for cover when they mention makeup or I shock a group of guys talking about such things as football, fishing, cars, construction, etc. by joining in and actually knowing what they are talking about."

Things Difficult for Trans-Heterosexuals To Do

LxMs, like LMs, are very competitive. But in this case, they tend to avoid physical combat, preferring the painting, photography, cabinetry or other intricate crafts. They tend to serve their wives openly, enjoying building and furnishing their homes to her preferences. They make excellent nurses. They also can show passive aggressive behavior, and can manifest a latent gothic orientation, as can LxFs (Chapter 9).

LxFs may also seek sexual dominance, and can have a dark side, for reasons to be discussed in later chapters. As suggested in the quote above, some LxFs are not comfortable with typical feminine stereotypes. They usually avoid much make-up, also sexy (hourglass figure accentuating) clothing, dresses, blouses with cleavage visible. They tend not to spend much time on their hair and avoid beauty parlor "hair-dos" beyond Page Boy types. As Mr. (Richard) Blackwell, originator of fashion's "Ten Worst Dressed Women List", once said: "Anyone can wear pants but only a woman can wear a dress"!

Observation #23: In Trans-Heterosexuals, the Emotion Side of the Brain is Reversed to the Left Side

One can compare which side of the brain is more emotional by using goggles designed only to permit peripheral vision on one side or the other (87, 110b). In most cases, the right side of the brain is the more emotionally reactive. However, it was noticed that for a significant percentage of the population, emotion was on the left side of the brain. These individuals were found to be the LxPs. Thus, interestingly, not only is their sexual identity reversed, but so is the emotional side of the brain. This may have a bearing upon the observation that LxPs appear to be emotionally cold. Such persons were present in a Swedish study investigating variation in the oxytocin receptor gene, which has been associated with emotional warmth and pair bonding (134a).

Observation #24: For Trans-Heterosexuals, Mirror Tracing Results are Opposite to Their Other Hemisity Test Results.

This was confirmed when it was found that trans-heterosexuals responded as LxPs for all biophysical and questionnaire tests except for one. In the Mirror Tracing Test, their left hand was fastest, and that test result had to be reversed to be congruent with the other hemisity tests (87). How this is related to other LxP hybrid properties is unknown. However, it adds weight to the list of differences in LxPs and further supports their yet to be recognized existence.

Familial Hybrid Interbreeding

Thus far, we have focused our attention upon hybrids between pure, wild type familial polarity parents, and ignored the much more common interbreeding between familial polarity hybrids. **Figure 14** illustrates the heterosexual offspring from the possible 16 hetero parental mating possibilities. In this figure, H for haremic (harem forming) replaces P for patripolar. And O for orgeic (orgy having) replaces M for matripolar.

In the first generation, the polarity of the children from the four possible matings is indicated within the boxes. Note that in the two native human subspecies, offspring are identical with parents, as indicated in the outer left and right boxes. However, in the R-R pair, their HO offspring are different from the OH offspring of the L-L pair.

In the second-generation crossed polarity breeding, there are 12 different hybrid parent couples involved. The possible children of these combinations include the hybrid combinations. Importantly, wild type throwback pure bloods also can result from all the hybrid-interbreeding combinations. Thus, a Bill Clinton, or a Steve Jobs can emerge from the general population.

Thus, it important to note that it is not possible to predict the random selection of hybrid offspring accurately. This

Figure 14. Heterosexual Offspring from the 16 F1 and F2 Generation Spousal Polarity Pairs

MATRIPOLARS
L-R pair (M-F)

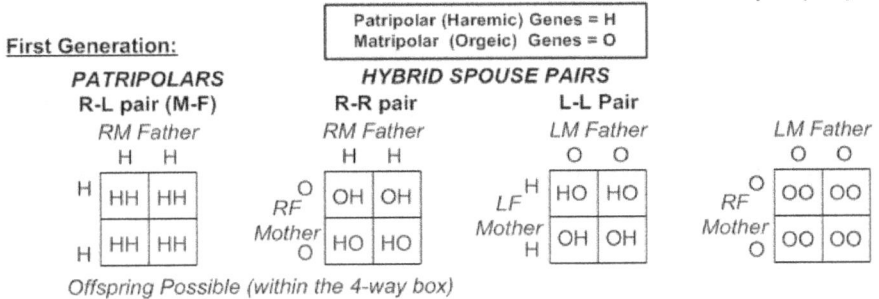

First Generation:

| Patripolar (Haremic) Genes = H |
| Matripolar (Orgeic) Genes = O |

PATRIPOLARS
R-L pair (M-F)

RM Father

	H	H
H	HH	HH
H	HH	HH

HYBRID SPOUSE PAIRS

R-R pair
RM Father

$_{RF}$ | O | H | H |
|---|---|---|
Mother O | OH | OH |
| HO | HO |

L-L Pair
LM Father

$_{LF}$	O	O
Mother	HO	HO
OH	OH

LM Father

$_{RF}$	O	O
Mother O	OO	OO
OO	OO

Offspring Possible (within the 4-way box)

Second Generation: (Only crossed polarity parentage outcomes are listed.)

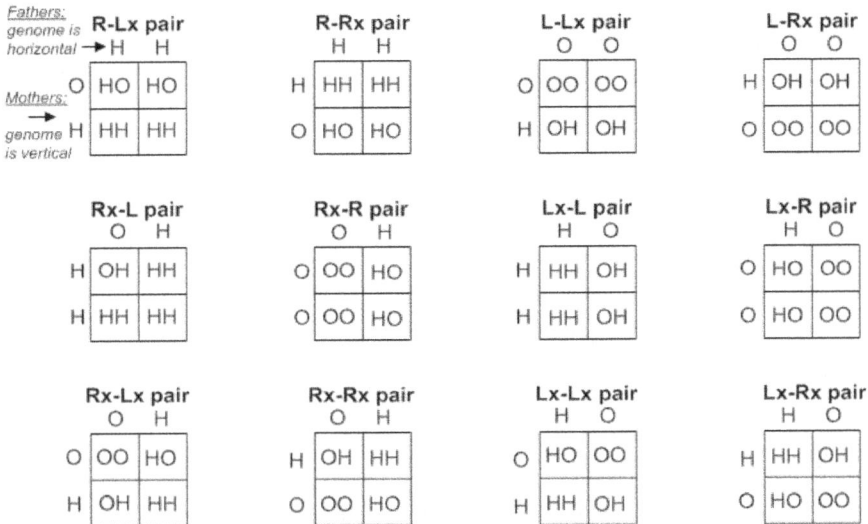

Fathers: genome is horizontal →
Mothers: → genome is vertical

R-Lx pair

	H	H
O	HO	HO
H	HH	HH

R-Rx pair

	H	H
H	HH	HH
O	HO	HO

L-Lx pair

	O	O
O	OO	OO
H	OH	OH

L-Rx pair

	O	O
H	OH	OH
O	OO	OO

Rx-L pair

	O	H
H	OH	HH
H	HH	HH

Rx-R pair

	O	H
O	OO	HO
O	OO	HO

Lx-L pair

	H	O
H	HH	OH
H	HH	OH

Lx-R pair

	H	O
O	HO	OO
O	HO	OO

Rx-Lx pair

	O	H
O	OO	HO
H	OH	HH

Rx-Rx pair

	O	H
H	OH	HH
O	OO	HO

Lx-Lx pair

	H	O
O	HO	OO
H	HH	OH

Lx-Rx pair

	H	O
H	HH	OH
O	HO	OO

Legend: The Eight Possible Heterosexual Offspring (within the 4-way Boxes above).:

H   H

| R-dominant King |
| L-supports Queen |

Patripolars

O   H

M | R-Dyslexic Chuckle-Head |
F | R-Dyslexic Ding-Bat |

Rx-Hybrids

H   O

M | L-trans-het Wimp |
F | L-trans-het Jock |

Lx-Hybrids

O   O

M | L-supports Prime Minister |
F | R-dominant Empress |

Matripolars

Thus, it is important to note that it is not possible to predict the random selection of hybrid offspring accurately. This book describes trends and is in no way absolute.

At the bottom of **Figure 14** is an arbitrary naming code to describe the eight possible heterosexual offspring of the four possible hybrid combinations in the boxes above. The pure patripolar male (RM) and female (LF) can be called the King and Queen. In contrast, the pure matripolar male (LM) and female (RF) can be called the Prime Minister and Empress. The male OH hybrid (RxM) is a cis-hetero dyslexic Chuckle Head, the female OH hybrid (LxF) is a trans-hetero Jock. The male HO (LxM) hybrid is a trans-hetero Wimp, while the female HO hybrid (RxF) is a cis-hetero dyslexic Dingbat. My good mother was one of those. Once when under pressure to take a family picture, she flashed a camera in her face.

Observation #25: LxPs Have the Potential To Be Bisexual

Both the wild type patripolars and matripolars have a double reinforcement to be strictly heterosexual in sexuality. This is because both their heterosexual personal sexual identity and their reproductive object focus enforce this orientation. In the next chapter on homosexuality, we will discuss the multiple effects of being reversed-wired regarding the gender of the person towards whom one feels affinity or love. What about everybody else? A significant percentage of people do not fit either of the types.

With trans-heterosexuals (effeminate men or masculinized women), the effect of reversal of one's sexual identity upon sexual behavior must also be addressed. Although LxPs are usually preferentially heterosexual, the cross-wiring of one's sexual identity reverses one of the pillars reinforcing heterosexual behavior. That is, from the perspective of one's internally reversed sexual identity, a heterosexual love object now should be homosexual one.

Thus, all LxPs have a latent bisexual element to their nature. They may experience sexual and emotional attractions and feeling for people of different genders at some point within their lives. This may never be acted upon or at the other extreme may result in pansexuality, where the gender of their everyday love object is irrelevant. This is strikingly different from normal or gay/lesbians for whom sex with the "wrong" gender is revolting.

There is no simple definition of bisexuality because bisexual people are a very diverse group. J. R. Little (77b) identifies at least 13 types of bisexuality, as defined by sexual desires and experiences. They are:
1. Alternating bisexuals:
> may have a relationship with a man, and then after that relationship ends, may choose a female partner for a subsequent relationship, and many go back to a male partner next.
2. Circumstantial bisexuals:
> primarily heterosexual, but will choose same sex partners only in situations where they have no access to other-sex partners, such as when in jail, in the military, or in a gender-segregated school.

3. Concurrent relationship bisexuals:

> have primary relationship with one gender only but have other casual or secondary relationships with people of another gender at the same time.

4. Conditional bisexuals:

> either straight or gay/lesbian, but will switch to a relationship with another gender for financial or career gain or for a specific purpose, such as young straight males who become gay prostitutes or lesbians who get married to men in order to gain acceptance from family members or to have children.

5. Emotional bisexuals:

> have intimate emotional relationships with both men and women, but only have sexual relationships with one gender.

6. Integrated bisexuals:

> have more than one primary relationship at the same time, one with a man and one with a woman.

7. Exploratory bisexuals:

> either straight or gay/lesbian, but have sex with another gender just to satisfy curiosity or "see what it's like."

8. Hedonistic bisexuals:

> primarily straight or gay/lesbian but will sometimes have sex with another gender primarily for fun or purely sexual satisfaction.

9. Recreational bisexuals:

> primarily heterosexual but engage in gay or lesbian sex only when under the influence of drugs and/or alcohol.

10. Isolated bisexuals:

> 100% straight or gay/lesbian now but has had at one or more sexual experience with another gender in the past.

11. Latent bisexuals:

> completely straight or gay lesbian in behavior but have strong desire for sex with another gender, but have never acted on it.

12. Motivational bisexuals:

> straight women who have sex with other women only because a male partner insists on it to titillate him.

13. Transitional bisexuals:

> temporarily identify as bisexual while in the process of moving from being straight to being gay or lesbian, or going from being gay or lesbian to being heterosexual.

Many of these people might not call themselves bisexual, but because they are attracted to and have relationships with both men and women, they are in fact bisexual. As would be expected from familial polarity, bisexuality is a third sexual orientation, separate from both heterosexuality and homosexuality. As illustrated in **Figure 15**, this has the potential for jealousy and family conflict.

It also has been noted that some natural elements of sexual behavior appear missing in LxPs. For example, in the wild type LF, the behavioral stereotype of kissing of the hand, followed by kissing of the arms, and then the neck, leads to the proffering of the breasts, followed by urging directed toward the genitalia. This motivational sequence appears to be missing in some LxFs.

While literally millions of people are bisexual, most keep their sexual orientation secret, so bisexual people as a group are nearly invisible in society. Gay men and lesbian women have long recognized the need to join together, create community, and to organize politically. Like a hybrid between races (heterosexuals and homosexuals) and not accepted by either, the bisexual has been particularly isolated and alone. Now they have

**Figure 15. A Woman Involved in a Potentially Dangerous
Bi-Sexual Relationship**

ASSOCIATED PRESS / J

A theater worker in Allahabad, India, removes a pos
Bollywood film "Girlfriend," based on a relationship be
two women, fearing protests over the movie's gay conte

been included into the LGBT (lesbian, gay, bisexual, and
trans-sexual) community.

The next chapter deals with the origin, meaning,
and genetic varieties of homosexuality.

CHAPTER 6. CIS- AND TRANS-HOMOSEXUALS: BORN THAT WAY

In the last two chapters, we discovered that for familial polarity cross breeding between RM-RF couples, their RxP hybrid children were dyslexic, while their LxPs children were trans-heterosexuals with reversed sexual identity and the potential for bisexuality. In this chapter we ask what type of hybrids might result from cross-breeding LM-LF cross-polar couples.

In the study of the hemisity of 406 couples, recall that about 20% of them were LM-LF couples. Of the hybrid children from these couples, an abnormal number were lost by miscarriage. Of the remainder who were born, there were more LPs than RPs. At least one quarter of them were recognized as homosexual. This is consistent with the roughly 10% of the global population who are self-identified homosexuals. The actual number appears to be as high as 30%.

Regarding familial polarity cross breeding, just as there are cis-heterosexual (CisHet) and trans-heterosexual (Trans-Het) offspring, so there are also cis-homosexual (Cis-Hom) and trans-homosexual (Trans-Hom) hybrid individuals. However as we will see, things are somewhat more complex here. This is illustrated in **Table 1**, the first geneologic map of homosexuality based upon familial polarity. In the first generation of LP and LxP parents, there are 16 different possible homosexual

Table 17. Types of Homosexual Offspring of L and Lx Hybrid Pairs (theory)

Parental Pairs: Male-Female	MALES: Common Description	Male-Bodied Homosex Offspring	Sexual Identity, Position	Sexual Identity, Position	Female-Bodied Homosex. Offspring	FEMALES: Common Description*
LM-LF	Gay Jocks	RyM	male	female	RyF	Lipstick Lesbians
	Twinks	LyM	male	female	LyF	Femmes
LM-LxF	Bears	RxzM	male	male	RxzF	Blue Jeans Lesbians
	Circuit Boys	LxzM	male	male	LxzF	Chapstic Lesbians
LxM-LF	Gay Listers	RzxM	female	female	RzxF	Diesel Dykes
	Art Fags	LzxM	female	female	LzxF	Butches
LxM-LxF	Drag Queens	RwM	female	male	RwF	Stone Butches
	Show Queens	LwM	female	male	LwF	Drag Kings

offspring, 4 of which are transhomosexuals. These, together with the LxM and LxF bisexuals (transheterosexuals), brings the sum to 18 possible types potentially practicing homosexuality. This creates quite a nomenclature challenge. Fortunately, the term LGBT (lesbian, gay, bisexual, or trans-sexual) does include all 18 types.

Observation #26: Cis-Homosexuals are Familial Polarity Hybrids

As shown in the top quarter of **Table 17**, there is only one cross-wiring abnormality in the four cis-homosexual offspring of pureblood parents. Such persons with a male body *and a male sexual identity* will powerfully be biologically oriented subconsciously to find and attempt to have sex with another male. Similarly, a person with a female body *and female sexual identity* will biologically be attracted to seek and have sex with another female.

However, as seen in the middle two quarters of the table, whether it was the father or the mother who was an LxP with reversed sexual identity, creates two sets of possible outcomes, just as in the earlier example of mules and hinneys. This influences the degree to which the sexual identity in these two sets of homosexual offspring is shifted. For example male bodies range from male-like Bears to female-like Art Faggs. For female bodies, they range from female-like Chapstick Lesbians to male-like Butches.

But, it goes further than this. Each homosexual also has the choice of forming a same sex partnership with a partner of the same or opposite sexual identity. This creates the possibility for a homosexual with a male identity to match with a male homosexual of female identity, creating complimentary top-bottom partnerships, rather than top-top or bottom-bottom non like-like pairs.

Beyond this, it is also possible to have RP dominant bottom partners matched with LP supportive top partners, or vice versa: RP dominant top partners with

LP bottoms. This is quite a large spectrum of non-fertile, only partially recognized match up possibilities. As a result, the complexities of life confronting a young homosexual can seem overwhelming.

Observation #27: Trans-Homosexuals Are Familial Polarity Hybrids

As shown in the bottom quarter of **Table 17**, the trans-homosexuals are the product of two cross-wiring anomalies. That is, for persons with male bodies, not only are they attracted to other males, but also their sexual identity becomes female. Similarly, for persons with female bodies, not only are they attracted to other females, but they also have a male sexual identity. In this case, as contrasted to all others, there is often a conscious identity crisis. The person feels trapped in the body of the wrong sex. This often leads to attempts at surgical remodeling, one famous case being Chaz Bono, a "male" *Dancing with the Stars* contestant in the 2011 season. More recently, the now female Caitlyn Jenner, former male Olympic star has been in the news.

Famous Homosexuals

Table 18 lists some of the many famous LGBTs, both today and in history. Clearly, being cross-wired for one's reproductive target in no way impairs general intelligence.

Biological Significance and Consequences of the Misdirected Motivation of Homosexuality

Reproduction is essential for the continuation of life as

Table 18. Notable Gay and Lesbian Individuals

Gays	Lesbians
Socrates	Sappho of Lesbos
Plato	Ruth, Biblical
Alexander the Great	Sister Benedetta Carlini
Julius Cesar	Christina, Queen of Sweden
Aristotle	Queen Anne
Pope Julius III	Marie Antoinette
Richard the Lion Heart	Mary Hamilton
Michelangelo	Eleanor Roosevelt
Leonardo Da Vinci	Emily Dickenson
Montezuma II	Gertrude Stein
William Shakespeare	Alice B. Toklas
Franz Schubert	Jane Addams, Nobel Peace Prize
Peter Tchaikovsky	Colette
Chief Crazy Horse	Marlene Dietrich
T. H. Lawrence	Drew Barrymore
Walt Whitman	Frida Kahlo
Henry Thoreau	Virginia Woolf
Cole Porter	Margaret Fuller
J. Edgar Hoover	Martina Navratilova
Alan Turing	Billy Jean King
T.H. Lawrence	Lily Tomlin
Rock Hudson	Jody Foster
Leonard Bernstein	Christina Aguilera
Greg Louganis	Wanda Skykes
Elton John	Rosie O'Donnell
Anderson Cooper	Ellen Degeneres

well as for the survival of the species. Thus, a drive to reproduce is deeply embedded in all life forms. It is one of the most incessant and powerful of all mammalian drives. Relatively speaking, the act of copulation appears to be one of the highest rewarded of all behaviors. Research indicates that in humans during their reproductive phase, the sex driven reproductive impulse occurs many times a day in both sexes. This drive is just as strong in homosexuals and pedophiles as in heterosexuals. However, the objects of their intense reproductive drive have been switched to non-fertile partners. In the case of homosexuals, those objects are adults of the same sex. If the feelings are not mutual, the person selected as the reproductive object can simply decline, as is common in nature. The same cannot be said if the reproductive object is a helpless child, as in pedophilia, the topic of the next chapter.

For some, second only to the rewards of copulation, are the joys of parenthood and family. This remains easily available in the trans-heterosexuals who remain quite fertile and usually form stable families. However in familial polarity hybrid homosexuals, a miswiring to an inappropriate reproductive target effectively compromises these pleasures. Not only does a lot of spurious same sex courtship with its highly rewarded frantic physical contact result, but millions are deprived of the "joys" of parenthood. Thus, the conversion of one's love object to a member of the same sex, causes more havoc than just the miswiring of one's own sexual identity.

Recently, homosexual couples of both sexes can and do form modern families with children. In the case of lesbians, this is made possible by the insemination of one of the partners. In the case of gay couples, adoption is possible, once it is assured that they are not also pedophiles: 21,740 same sex couples adopted children in 2009, up from 6,477 in 2000.

Observation #28: Homosexuals Can Also Be Dyslexic

It is interesting that among the possible male homosexuals of the male RyM types, two public figures CNNs Anderson Cooper and Actor Robin Williams actually appear to be dyslexic RxyMs, while amazing Olympic diver, Greg Louganis, another dyslexic homosexual, appears to be an RxzM. Then there is the case of Olympic track star Bruce/Caitlyn Jenner, one of the most famous transgender celebrities who is also dyslexic.

Other Problems Homosexuals Face

Social behavior is very competitive with highly evolved strictly defined boundaries, especially regarding courtship, reproduction, and family rearing. Those operating outside of these boundaries are automatically subjected to criticism, rejection, and ostracism and homophobic attack.

Young isolated homosexuals often opt to keep their mis-wired inclinations to themselves, only acting them out in secret. This is, of course called "staying in the closet". Recently, homosexuals have been seeking to change the social boundaries, by indicating that, aside from their like-like sexual preferences, they are normal. Many high-achieving or wealthy homosexuals have been coming out of the closet, some of them in the mass media. This is actually effecting a change in social standards, perhaps for the good.

However, this social accommodation is not yet fully accomplished. Thus, most homosexual children, often born to a heterosexual couple have the incredible, often overwhelming challenge of figuring out who they are when social norms differ from their unnatural inclinations. Those with reversed sexual identities often find that they cannot avoid the expression of their reversed gender, both by word inflections and body language. Peers spot trans-sexuals immediately and naturally reject, exclude, and attack them. The resulting bullying is a problem within schools internationally and the subject of intensive research. When they find they cannot change their nature, it can create despair within the homosexual child. Without a network with similar peers, it appears as if fate was forever against them. When several of them find themselves to be in the same boat, things can go bad, as in the Columbine Massacre. Not recognizing that a positive outcome might be possible, often leads to the dark side. This topic will be taken up in Chapter 8.

As if this were not enough, the natural promiscuity of gays has attracted sexually transmitted parasites, most horribly the HIV virus responsible for global AIDs epidemic and the killing of millions. It is no fun being a familial polarity hybrid.

In some cultures, trans-homosexuals have sought relief from persecution by recreating themselves as a legitimate "third sex" with large memberships.

Familial polarity makes irrelevant the question of why homosexuality, as the result of mutations which do not enhance the survival of the species, could exist in a Darwinian world. Homosexuality is not caused by mutated genes, but rather results from the genetic mismatch occurring from the cross-breeding between the two human species. There are no "gay" genes.

CHAPTER 7. PEDOPHILES: BORN TO SUFFER OR BE HATED

Definition of Pedophilia:

According to the American Psychiatric Association's Diagnostic and Statistical Manual, IV TR, pedophilia arises from the recurrent, intense, sexual urges towards and sexual fantasies about prepubescent children and on which feelings they have either acted or which cause distress or interpersonal difficulty. It includes incest or domestic sexual abuse directed against children.

Rate of Occurrence of Pedophilia:

This is about 3-9% of the population (about 1 in 20). However, pedophiles have sexually molested a much larger percent of American children, an estimated 20% of them, making pedophilia the most common of the sexual deviations.

Types of Pedophiles

Pedophiles are sexually obsessed with children and their goal is to get close to children of their preferred age range. Pedophiles have obsessive-compulsive characteristics, and they live within the sexual fantasies they create. Some actually believe that children want molestation and that they enjoy it. However, most know, deep down, that what they are doing is wrong… they just can't stop.

Pedophiles can be married, single, male, female, bisexual, heterosexual, or homosexual. There are far more heterosexuals pedophiles than homosexuals. In fact, the ratio is 11:1.

No matter what their sexual preferences are, pedophiles are often also attracted to adults. A pedophile that is attracted to both children and adults is called a Regressed Offender because they bounce back and forth between an adult sexual relationship and criminally assaulting children.

There are two main types of pedophiles: they are the situational pedophiles and the preferential pedophiles. Situational pedophiles will go after any group that is defenseless, such as the elderly, the mentally challenged, the handicapped, etc. This type prefers children, but goes after another group if they feel stressed. There are three subtypes of Situational Pedophiles: The Regressed Offender Pedophiles, Indiscriminate Pedophiles, and Immature Pedophiles.

The second type, the preferential pedophiles like children of a distinct age group, and they typically do not deviate from this age group. There are two subtypes: The Seductive Pedophiles, and the Sadistic Pedophiles.

Observation #29: Cause of Pedophilia: Sexual Drive Becomes Miss-Wired to an Immature Reproductive Target

Thus, far in our investigation of the effect of Familial Polarity Crossbreeding to produce hybrid offpring, we have dealt with its effects upon the first and second-generation children to produce dyslexics, trans-heterosexuals, and cis- and trans-homosexuals. In particular, we have seen that in each of these anomalies, an error in CNS crossings occurs to wreck havoc upon a specific process, in these cases of memory access, sexual identity, or mate preference (relevant to pedophilia).

The Hybrid Familial Polarity Pedophilia Hypothesis

It takes little imagination to recognize the potential for third-generation hybrid matings to produce additional deviations. These include further anomalies of sexual partner preference, including the cross-wired misdirected selection of children as a reproductive sexual target, diagrammed in **Figure 16**.

Specifically, the "familial polarity pedophilia hypothesis" asserts that cross-breeding of a specific, yet unknown pair of second or third-generation familial polarity hybrids will result in pedophile offspring whose natural sexual object choice will be cross-wired to children.

The few known pedophiles whose hemisity has been determined were left brain-oriented males. Such tended to show abnormal psychosexual arousal to child pornography.

Pedophilia Is Irreversible and Incurable

Like with dyslexia and homosexuality, the CNS cross wiring of pedophilia is presently irreversible. Throughout each day, like the rest of us, pedophiles are stimulated to act on their reproductive and pleasure drive to have sex. They have proven very creative in developing situations, even institutions, bringing them fresh child victims. They may be married, and the marriage may come across as a happy, successful union. It is of interest, however that a number of adult males demand that their

Figure 16.
The Familial Polarity Pedophilia Hypothesis

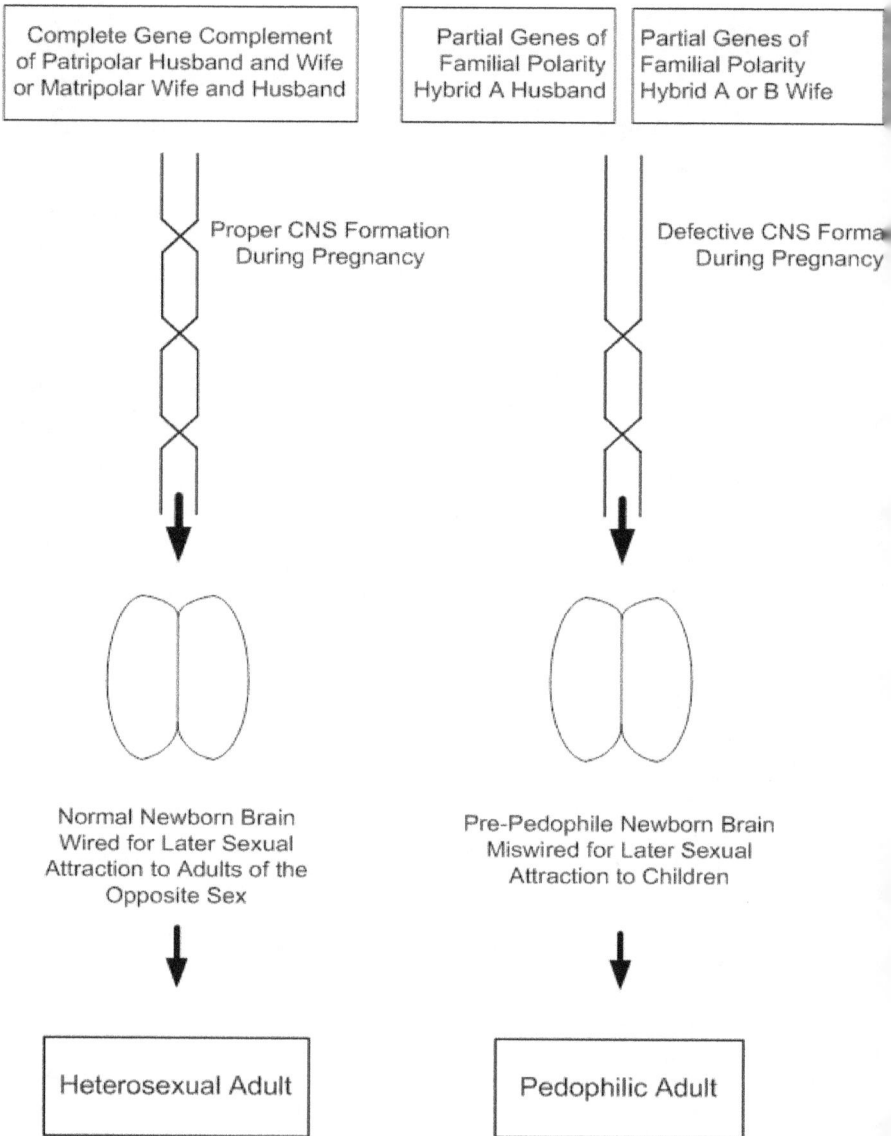

| Complete Gene Complement of Patripolar Husband and Wife or Matripolar Wife and Husband | Partial Genes of Familial Polarity Hybrid A Husband | Partial Genes of Familial Polarity Hybrid A or B Wife |

Proper CNS Formation
During Pregnancy

Defective CNS Forma
During Pregnancy

Normal Newborn Brain
Wired for Later Sexual
Attraction to Adults of the
Opposite Sex

Pre-Pedophile Newborn Brain
Miswired for Later Sexual
Attraction to Children

Heterosexual Adult

Pedophilic Adult

girlfriends and wives have their pubic hair removed, perhaps to imagine they are having sex with a pre-pubic child. The molester may even love his wife, but he uses his role as the happy family man to build trust and gain access to children. Pedophiles often have numerous victims and many claim to have abused hundreds or even thousands of children over their careers.

Infiltration of Child Service Institutions by Pedophiles

Pedophiles will make every possible effort to be around children. They choose professions that will allow them to interact with children in their preferred age range. Then often form networks with other pedophiles and share child pornography. Child molesters have taken positions as youth pastors, boy-scout leaders, coaches, teachers, child-care providers, children's choir directors, nurses, counselors, and employees at establishments that serve children.

They slowly get to know their targets, making friends with the child and starting the "seduction" process. They try to become the child's confidant and often offer email addresses, home addresses, and phone numbers. Child molesters remain patient and often have several children they are trying to seduce at the same time. This type of seduction is all about gaining trust with the child. Children are not sexually attracted to their abusers, but they do start to trust them as a friend.

These molesters get to know the child so well that when they do finally molest them they have information to blackmail them. For example, "Don't you tell or I'll tell your parents this…" They get to know their victims so well that they know how to manipulate them and confuse them. They essentially become stalkers, but they are not obvious. Pedophiles usually try to establish a good reputation in their communities. Once they have gained this reputation, people do not really pay attention because they are trusted. This is why the seduction and stalking goes unnoticed.

Pedophiles are narcissistic by nature and their bragging usually comes across as narcissistic. Many pedophiles tend to like children of a certain gender and they typically do not deviate from their preferred age range. Many prefer girls that are too young to get pregnant. The younger the girls are, the less chance there is of vaginal infections. Girls are now reaching puberty at much younger ages, some as early as the 4th grade. "Eight is too late" is a phrase molesters often use.

Pedophiles like to collect things that relate to their sexual perversions. Some examples include pornographic pictures, pornographic videos, and books on sexuality. These items are often introduced to future victims. This interaction can later be used as blackmail, i.e., "I will tell your parents" or they can use it to confuse the child. Child molesters could say, "Well you acted like you were interested in… so I just tried to show you… it is your fault." Other items include sex toys, photographic equipment, items for a private darkroom, and any items that children, in the preferred age range, would relate to.

Pedophilic Women

McConaghy estimates a 10 to 1 ratio of male-to-female child molesters (80b). It is believed that the true number of female pedophiles is underrepresented by this estimate. The reasons for this may include a "societal tendency to dismiss the negative impact of sexual relationships between young boys and adult women, as well as women's greater access to very young children who cannot report their abuse, among other explanations."

The slight amount of attention given to female pedophiles is staggering. According to David Finkelhor's (37) numerous studies, women's sexual abuse of children is much more serious than men's because women are more likely to have abused more children for a longer period of time. They are more intrusive and are more likely to use higher rates of force than men do. Finkelhor found that women molested more than 60% of the total children who were molested. There are three categories of female pedophiles: 1. Women Coerced to Molest by their Husbands, 2. Women Seeking Adolescents, and 3. Women Who Hurt Little Kids.

Usually these women's victims are their own children. When these offenders see their victims, they see themselves... and hate what they see. She sees what happened to herself as a child. She sees the results of all the horrible things she's experienced... and is now forcing on her child. Many times, it is her unwanted child. But why wouldn't she want her own kid? Here's why: incest, rape, or the reason can be something as shallow as "Your dad left me because you were a girl and not a boy."

When the child is disobedient, just as any child can bes, the female pedophile sees it as deliberate maliciousness or rebellion. Sexual abuse becomes a means of revenge or punishment. In all cases, women in this category are very sadistic; they get pleasure from their victim's pain. Thus, these mothers could have sons who become serial killers of women because of the residual deep hatred of her.

Consequences of Pedophilia Upon Children

Amazingly, in spite of many reports of lasting sexual dysfunction among victims (37), there still exists the ancient Greek controversy as to whether the effect of sexual penetration of pre-adolescent boys or girls by pedophiles is good or bad.

Given the complex stratagems that pedophiles employ, it is understandable that few victims realize the ultimate goal of their newfound "friend." A U.S. Department of Justice report on child sexual exploitation states (113b), "Most preferential sex offenders spend their entire lives attempting to convince themselves and others that they are not perverts and that they love and nurture children. The tragic fact is that children who are being preyed upon "have been carefully seduced and often do not realize they are victims. They repeatedly and voluntarily return to the offender. The truth often does not become apparent to them until after they have been abandoned: Then they see that all the attention, affection, and gifts were just part of a master plan to use and exploit them."

According to Finkelhor (36), certain conditions must prevail for informed consent to occur. The person must understand what he or she is consenting to, and must have true freedom to say yes or no. Are children capable of fulfilling this condition in "consenting" to sex with adults? Finkelhor denies that they can.

For one thing, he says children lack the information necessary to make an 'informed' decision about the matter. They are ignorant about sex and sexual relationships. More important, they are generally unaware of the social meanings of sexuality. For example, they are unlikely to be aware of the rules and regulations surrounding sexual intimacy or what it is supposed to signify. They are uninformed and inexperienced about what criteria to use in judging the acceptability of a sexual partner. They cannot know what likely consequences it will have for them in the future.

Children may genuinely like the adult who is molesting them. Or more to the point, may have become emotionally or otherwise dependent upon the pedophile. They may willingly spend time with their molester and may even find some enjoyment in the physical sensation of pleasure. But all this is not enough: The fundamental conditions of genuine consent are not present. Children "lack the knowledge the adult has about sex and about what they are undertaking... In this sense, a child cannot give informed consent to sex with an adult.

Effects of Pedophilia Upon the Population

Victimization by pedophiles has negative early and long-term effects upon children (37). In addition, high-profile media attention to pedophilia has led to incidents of moral panic, particularly following reports of pedophilia associated with satanic ritual abuse and day care sex abuse. Instances of vigilantism have also been reported in response to public attention on convicted or suspected child sex offenders. For example in 2000, following a media campaign of "naming and shaming" suspected pedophiles in the UK, hundreds of residents took to the streets in protest against suspected pedophiles. These eventually escalated to violent conduct requiring police intervention

Protection from Pedophiles, Child Molesters, and Sexual Offenders

Pedophiles, child molesters and sexual offenders are names that apply to those that stalk and molest our children. They are dangerous and often invisible. One can take the usual precautions to protect your children from the unknown child molesters, but you also have a great resource for convicted child molesters. In the US, The Registered Offender List is a list of every convicted sexual offender in your city. After entering your address, you will receive detailed photos of the child molester, appearance details, a neighborhood map, street address, aliases, and the complete details of the molester's conviction. Parents can also sign up to be notified when a child molester moves into their area.

Significance of the "Pedophiles are Familial Polarity Hybrids" (PFPH) Hypothesis .

Until the present, the origin of pedophilia has remained unknown. Personality problems may be evident, and the pedophile often shows little or no concern for the effects of his sexual behavior on the child. Certain researchers have reported that psychotherapy in conjunction with the use of testosterone-lowering drugs (chemical castration) has substantially reduced the libido in certain male pedophiles. However, this is not believed to be effective for power seeking or sadistic pedophiles.

A few researchers attribute pedophilia along with the other sexual deviations to biology. For example, that abnormally elevated in utero levels of testosterone, one of the male sex hormones, is thought to predispose men to develop deviant sexual behaviors.

As would be predicted from my PFPH hypothesis, no researchers have claimed to have discovered or mapped any gene for pedophilia. I assert that there are no pedophilia genes, just the mismatched genes of cross-polar hybrids.

Most experts regard pedophilia as resulting from psychosocial factors rather than biological characteristics. Some think that pedophilia is the result of having been sexually abused as a child, or from some of the person's interactions with parents during their early years of life. Other researchers attribute pedophilia to arrested emotional development; that is, the pedophile is attracted to children because he or she has never matured psychologically. Some regard pedophilia as the result of a distorted need to dominate a sexual partner. Since children are smaller and usually weaker than adults are, they may be regarded as nonthreatening potential partners.

The present proposal that *pedophiles are family polarity hybrids with miswired sex object preference*, is the first potentially testable hypothesis explaining the origin and mechanism of pedophilia. It is my prediction

that the determination of the hemisity of many pedophiles, along with their family polarity lineages, will provide more specific details regarding its detection, prediction, and prevention.

Table 18b. Important Pedophiles

Cheng I: Hong Kong pirate
Muhammed and 6 yr old wife Aishah
Ghandi
William S. Burroughs
Pope Julius III

James I of England
Pyotr Ilyich Tchaikovsky
Jules Verne
T.E. Lawrence
Joseph Smith

Socrates
Plato
Michael Angelo
Shakespeare
Andre Gilde, Nobel in Literature
Oscar Wilde

THE TWO HUMAN SPECIES: FAMILIAL POLARITY

CHAPTER 8. SCHIZOPHRENIA UNLOCKED

The Enigma of Schizophrenia

Schizophrenia has been called the "worst disease affecting mankind". This is because of the extent of the loss of abilities that it causes and its usually life-long course. It is among the top ten leading causes of disease-related disability in the world. However, despite vigorous study over the past century, its origin and mechanism remain obscure, and available treatments are only modestly effective.

At first, the progress of schizophrenia research was retarded because it was thought to be a psychological illness. Yet, our present ignorance of the nature of schizophrenia cannot be due to a lack of later neuroscience research findings. In fact, the several hundreds of thousands of publications to date have described thousands of discrete findings. However, the question remains, what do these facts tell us about the nature of schizophrenia? It remains a disease whose mechanism is totally unknown.

Seemingly Normal People Are Stricken in Young Adulthood

An important and often overlooked fact about this mentally disabling disease is that it strikes people in their early adulthood. That is, after having lived essentially normal lives for 17-28 years, schizophrenia is activated over a period of up to 30 months to cause its victims literally to go crazy. In general, males are attacked earlier than females, with 1.4 males falling ill per female. The global frequency of schizophrenia is almost 1% overall.

Symptoms of Schizophrenia:

The following are descriptions taken from the National Institutes of Mental Health website. There are positive, negative, and cognitive symptoms.

Positive symptoms are psychotic behaviors not seen in healthy people. People with positive symptoms often "lose touch" with reality. These symptoms can come and go. They include:

Hallucinations These are things a person sees, hears, smells, or feels that no one else can see, hear, smell, or feel. "Voices" are the most common type of hallucination in schizophrenia. Many people with the disorder hear voices. The voices may talk to the person about his or her behavior, may order the person to do things, or warn the person of danger. Sometimes the voices talk to each other.

Delusions These are false beliefs that are not part of the person's culture and which greatly resist change. The person believes delusions even after other people prove that their beliefs are not true or logical. People with schizophrenia can have delusions that seem bizarre, such as believing that neighbors can control their behavior with magnetic waves. They may also believe that people on television are directing special messages to them, or that radio stations are broadcasting their thoughts aloud to others. Sometimes they believe they are someone else, such as a famous historical figure. They may have paranoid delusions and believe that others are trying to harm them, such as by cheating, harassing, poisoning, spying on, or plotting against them or the people they care about. These beliefs are called "delusions of persecution."

Thought disorders These are unusual or dysfunctional ways of thinking. One form of thought disorder is called "disorganized thinking." This is when a person has trouble organizing his or her thoughts or connecting them logically. They may talk in a garbled way that is hard to understand. Another form is called "thought blocking." This is when a person stops speaking abruptly in the middle of a thought. When asked why he or she stopped talking, the person may say that it felt as if the thought had been taken out of his or her head. Finally, a person with a thought disorder might make up meaningless words, or "neologisms."

Movement disorders These may appear as agitated body movements. A person with a movement disorder may repeat certain motions over and over. In the other extreme, a person may become catatonic. Catatonia is a state in which a person does not move and does not respond to others.

**Negative symptoms** are associated with disruptions to normal emotions and behaviors. These symptoms are harder to recognize as part of the disorder and can be mistaken for adolescence, depression, or other conditions. These symptoms include the following:

- **"Flat affect"** where a person's face does not move or he or she talks in a dull or monotonous voice.

- **Lack of pleasure** in everyday life
- **Lack of ability to accomplish** The inability to begin and sustain planned activities
- **Speaking little**, even when forced to interact.

People with negative symptoms need help with everyday tasks. They often neglect basic personal hygiene. This may make them seem lazy or unwilling to help themselves, but the problems are symptoms caused by the schizophrenia.

**Cognitive symptoms** are subtle. Like negative symptoms, cognitive symptoms may be difficult to recognize as part of the disorder. Often, they are detected only when other tests are performed. Cognitive symptoms include the following:

- **Poor "executive functioning"** (the inability to understand information and use it to make decisions)
- **Trouble focusing** or paying attention
- **Problems with "working memory"** (the inability to use information immediately after learning it).

Cognitive symptoms often make it hard to lead a normal life and earn a living. They can cause great emotional distress.

Failure of Genetic Studies: Due to the Lack of Proper Context

Sixty percent of schizophrenics have no relatives with the disease. But, in cousin marrying cultures, such as in certain Islamic and third-world populations, its prevalence is more than twice (0.7%) that of non consanguineous marriages in first world cultures (0.3%). However, the role of suggested genetic factors appears to be a limited one. In affected monozygotic pairs in a large sample, the other twin had no schizophrenia 85% of the time.

The largest most comprehensive genetic study of its kind, involving tests of several hundred single nucleotide polymorphisms (SNPs), was performed with nearly 1,900 individuals with schizophrenia or schizoaffective disorder and 2,000 comparison subjects. Its findings were reported in 2008 (104). There was no evidence of any significant association between the disorders and any of 14 previously identified SNPs. The statistical distributions suggested nothing more than chance variation.

Human geneticists in general have been logically trapped in the "mutated gene" context. However, although genetic in origin, mutated genes do not cause schizophrenia. As we shall see, a paradigm shift is needed. Genetic mismatch in utero occurs in hybrids between two different human species. No genetic mutation cause schizophrenia. There are no schizo-phrenia genes.

Baffling Changes in Brain Structure and Activity:

First, in terms of neuroanatomy, there appears to be a normal phase of neuronal pruning in the young adults leading to a natural slight reduction in total brain volume. However, in the case of the schizophrenic, this pruning is excessive, leading to the abnormal enlargement of some ventricular spaces due to loss of essential brain tissue (112).

Abnormal loss of brain volume occurs in the frontal cortex, in limbic areas (located deep in the brain), which contain the amygdale (of emotions), and the hippocampus (of memory). It has been noted people with schizophrenia have smaller left hippocampi, which is associated with problems in verbal and other memory. Critically, structures of the striatum (required for same-different matching comparisons) are also reduced in size.

Second, in terms of regional brain activity, as revealed by functional MRI, there is decreased activity in the prefrontal cortex, both at rest and in cognitive challenge studies. Abnormal activation patterns are also seen in several other brain regions during performance of various cognitive tasks. For example, the anterior cingulate cortex executive system fails to activate under cognitive challenge.

Third, neurochemically, overproduction of the neurotransmitter dopamine and underproduction of the neurotransmitter glutamate occur. Thus, dopamine activators, such as amphetamine, worsen symptoms. In contrast, the dopamine D2 receptor blocker drugs, now given to schizophrenics, alleviate them. Correspondingly, the glutamate receptor activator, ketamine can produce schizoid symptoms even in normal people. In addition, the debilitating presence of schizophrenia is exceedingly stressful to its victims. This causes the elevation of the potent stress hormone CRF (Corticotropin Releasing Factor) to cause terror. Intracerebral injection of CRF caused agitated demands to commit suicide.

What Normally is Happening in the Brain in Early Adulthood When Schizophrenia Strikes?

A Wisconsin Law Review report by Callum, (15) reviewed information relating post-adolescent behavior and cognitive development. Historically, scientists believed that the human brain ceased development when an individual reached the age of 12. With the advent of anatomical and functional magnetic resonance imaging (MRI and fMRI), scientists have found evidence that the brain continues to develop through adolescence and even through "emerging adulthood". This period includes identity exploration based on "love, work, and world views." It was also during this period that the author of the present book first became able to understand quantitative relationships and to perform mathematics.

In 1999, the results of a large study were published using MRI to investigate the growth and development of the human brain during the period of 4-20 years (48). Significantly, it was found that the cerebellum was the last to develop, its growth period extended into the 20s, long after other regions have ceased growing. Most areas of the brain are highly heritable in size, but the cerebellum was not. It is unknown what determines its size. This size maturation during the time window when schizophrenia strikes further supports the author's suggestion of possible cerebellar involvement in the disease (87).

fMRI has permitted researchers to determine which regions of the brain are activated by questions posed or in scenarios suggested to an individual inside the MRI scanner. For example, in one study on adolescents and adults, researchers found that, when the groups viewed pictures of facial expressions, their brain activity patterns were very different. Adults correctly identified

the facial expressions by relying on the prefrontal cortex-
the area of the brain involved in judgment, reason, and
planning. Adolescents, however, found it difficult to determine correct responses. The adolescents relied mostly
on the amygdalar region of the brain associated with gut
reactions, instinct, and overall emotional responses. As
the teens aged, they came to rely more on the prefrontal
cortex and less on the amygdala.

This late maturation of brain and behavior was related to a process called myelination that occurs when
non-neuronal support cells wrap themselves around the
active neuronal tract to give it insulation and thus more
efficiency. The effect of this process turns a dirt road into
a superhighway, so to speak. This is associated with increases in "white matter" density. Scientists have associated differences in myelination with varying levels of
cognitive maturation. Research has shown that white-
matter maturation, particularly in the frontal lobe of the
brain (which includes the prefrontal cortex), correlates
with measures of executive function.

While the brain is forming, it produces more cells
and connections than will eventually be needed. During
childhood, the brain undergoes a "pruning" process in
which unneeded brain cells and connections are eliminat-
ed. Although the human brain is 95 % of its adult size before a child reaches the age of six, its development is far
from over. The brain experiences yet another pruning period and increased myelination during adolescence.

Overall, research has shown that gray-matter volume has the following three part developmental trajecto-
ry: its volume within the brain increases during child-

hood, peaks at adolescence, and decreases in both late adolescence and young adulthood. Evidence shows that the prefrontal cortex does not mature fully until the mid-20s, and that myelination continues throughout the 20s.

Myelination generally occurs from back of the brain to front, thus the prefrontal cortex, located just behind the forehead, is the last part of the brain to mature. It acts as the "CEO" of the brain, using controlling planning, working memory, organization, and modulating mood. In other words, the final part of the brain to grow up is the part capable of deciding, *I'll finish my homework and* then *I'll instant message my friends about seeing a movie.* Because the prefrontal cortex governs impulsivity, judgment, planning for the future, and foresight of consequences, it is the source of the very characteristics that may make one responsible for their actions.

Why Do Schizophrenic Symptoms Occur in Early Adulthood?

The Uterine Pathology Idea: Ironically, the apparent absence of mutated genes as a cause for schizophrenia brings us to the pathogenic theory of schizophrenia. Here, it is thought that in certain cases of schizophrenia, there is an interaction of the developing fetus with pathogens, such as viruses, or with antibodies from the mother created in response to these pathogens. Substantial research suggests that exposure to certain illnesses (e.g., influenza) in the mother of the neonate (especially at the end of the second trimester) causes defects in neural development that may emerge as a predisposition to schizophrenia just after puberty, as the brain develops to its final form. In suggestive support of a viral infection, 10% more children born in Europe in March (spring) developed schizophrenia than those born in September (fall).

In later versions of the uterine pathology model, it is felt that early developmental disruptions somehow miswire a brain in such a way that during brain reorganization steps occurring late adolescence or early adulthood, schizophrenia strikes and reality breaks down.

More specifically, there are ideas about the neural subplate, a structure required to guide other neurons to their proper sites. It is formed at about the fourth month of pregnancy. It gradually disappears almost entirely within the first month of postnatal life, having performed its task of aligning neurons toward their proper targets. The migration of brain cells through the neural subplate occurs almost entirely in the second trimester of fetal development in theory. If this migration is somehow disrupted, cells may end up in the wrong place or have faulty connections, such as found in schizophrenic patients. The mal distribution of these cells suggests an abnormality in the subplate function during pregnancy, as if there were a disconnectivity (i.e., miswiring) stemming from disturbed neuronal migration in the second trimester *in utero* causing neuronal misrouting in the schizophrenic. Clearly, the stage is set for schizophrenia years before the actual descent into psychosis, even if there are not overt symptoms.

The Destructive Brain Firestorm in Young Adult Schizophrenics

If schizophrenia is rooted in the first months of a person's life, why do decades pass before psychotic symptoms appear? Schizophrenia has been called a three-act tragedy. In the first act, prenatal risk factors set the story in motion. In act two, subtle coordination and social problems generally go undetected by parents and teachers. And then, in the final act, there is a catastrophic and seemingly sudden lurch into mental illness in the late teens or early 20s. In schizophrenia, for some reason an aberrant and excessive wave of neuronal pruning (3b) occurs at adult onset . This causes the reported abnormal tissue loss from back to front brain enlarging the ventricle spaces. Strong evidence suggests that schizophrenia also involves decreased communication between the left and right sides of the brain (147).

Survival With a Broken Brain: The Sufferings of the Schizophrenic

The absence of memory can be a very disabling thing. This condition can be approximated in normal people by a high dose of cannabis. Without memory, one cannot talk meaningfully because one does not remember what the topic of conversation was, nor the comment to which one is responding, nor even the content of the sentence fragment one has already begun to utter. In the mean time, inner activity is so rapid that the actions in the external world seems exceedingly slow. As a passenger in a car in town, it seems hours before the next block comes by. As a driver, one forgets one's destination, and can become lost, and even disorientated as to lane.

Leaving the topic of memory for the moment, a second failure that occurs in the thinking of the schizophrenic is called loosening of association. This actually results from failure of the striatal same- difference-matching system in the brain core. Matching is essential to tell whether two things are the same or different. Matching skills enable us to tell the difference in sizes and amounts. It kept smaller reptiles from challenging larger ones and being killed. Matching of blood levels of glucose to the body standard enables us to avoid major over or undershoots occurring when we eat or are fasting. Matching is also required to recognize the many unconscious *contexts* that exist in life. For example, in the case of "stocks and bonds", are we talking about Wall Street or Pilgrim restraints?

Schizophrenia is an especially challenging disorder because, through matching failure, it disrupts the basic human ability to differentiate reality from fantasy. The schizophrenic loses the ability to maintain a stable context. The individual may shift from one topic to another completely unrelated topic without realizing they are making no logical sense. This is context loosening due to loss of matching ability.

This sometimes also happens to normal people under the influence of hallucinogens. One is talking about something, but that reminds one of another thing, which reminds one of another thing, etc, etc, each being a new context. In its extreme form, it produces a so-called "word salad" of incoherence from lack of logical connection. On the other hand, only a slight loosening of association can create a form of "genius", where one's thinking is outside of the box of the conventional context. Then, unusual contents are juxtaposed together into new and interesting creative syntheses.

But, with deepening schizophrenia, one cannot choose or control the wild spin of unmatched context. As one schizophrenic said, "If I am reading, I may suddenly get bogged down on a word. It may be any word, even a simple word that I know well. When this happens I can't get past it. It is as if I am being hypnotized by it. It's as if I am seeing the word for the first time and in a different way from any anyone else. It's not so much that I am absorbed in it, it's more like it is absorbing me." Thus, a context captures you.

And, since you are totally "at effect", you can only conclude that the context that captured you must be especially significant. As you endlessly begin to try to create some kind of sense of it, you become delusional. It is the only thing that makes sense to you. Motivation by normal rewards becomes impaired as your cost-benefit system weakens.

Having both memory problems and context problems, one can understand why schizophrenics find difficulty in getting anything done. Endless loops of incompletion entrap them. There may even be a lack of clear sense of where their bodies end and the rest of the world begins. Thus, they often cannot care for themselves. On the street, I have seen men with pants too large so that they must hold them up. If the hands are needed to eat or for something else, down go the pants... oops! They cannot plan ahead, even enough to find a rope to tie around their waists. The untreated schizophrenic it truly incapacitated.

Add to this a modicum of self-consciousness. That is, you know that you are failing to function and are acutely aware that this is obvious to others. They really can read your mind because nothing is in there. You naturally feel paranoia, because with no context stability or memory, in contrast, you can't understand their minds. Further, you are hungry, exposed, without shelter, clothing, or meaningful human contact. Nobody can match your contexts. With no context in common with others, you are, by definition crazy. You fear loss of control or victimization. You may not survive. This activates your stress response. Terror strikes. Talk about suffering! About 5% of schizophrenics do manage to kill themselves. Others attempt to accept the unfairness of life without completely giving in to it.

This doesn't even mention hallucinations and voices. Psychosis is said to be a secondary symptom, like fever is to an infection. If you can't think clearly, then you have delusions, hallucinations, ideas of reference, ideas of influence, and thought disorganization. Defects in working memory and loosening in associations are the core elements of this illness. Inner superego-like voices provide some release. At first they are comforting, then condemning, then conquering and finally controlling. These inner voices comment on a person's behavior, perhaps insulting them or sometimes giving commands.

Observation #31: Hybrid Familial Polarity Schizophrenia Model

So the stage is set: An unknown event appears to occur in the second trimester of pregnancy whose effects lay silent until they powerfully disrupt the last stages of development of the brain in young adulthood to produce the excessively pruned brain of the schizophrenic with all its symptoms. Although, in utero viral or inter-uterine other insults have been intensively sought, none been detected. Similarly, no defective genes have been found causing schizophrenia. This is because schizopherenia is not caused by a genetic mutation. I assert that it comes from cross species breeding genetic mismatches not schizophrenic genes.

The hybrid family polarity schizophrenia model, **Figure 17,** supplies the mechanism that answers these mysteries. Differences between the genes of interbreeding <u>Homo sapiens patripolaris</u> and <u>Homo sapiens matripolaris</u> create a mismatch deficit in the fetus of certain of their hybrids. Needed CNS factors for midline crossings or other fetal events are absent or in excess in the gametes of the hybrid offspring of the specific, yet to be identified hybrid-hybrid cross breeding pair. This leads to a defect in the formation or function of one of the neural subplate, perhaps in the corpus callosum, in the second trimester of pregnancy. Years later, during the activation of the little known early adulthood brain maturations steps, this abnormality results in the improper activation or specificity of neural pruning. The cerebellum is also implicated. Over a period of months, this leads to massive tissue losses of the frontal cortex, hippocampus,

Figure 17

The Familial Polarity Schizophrenia Hypothesis

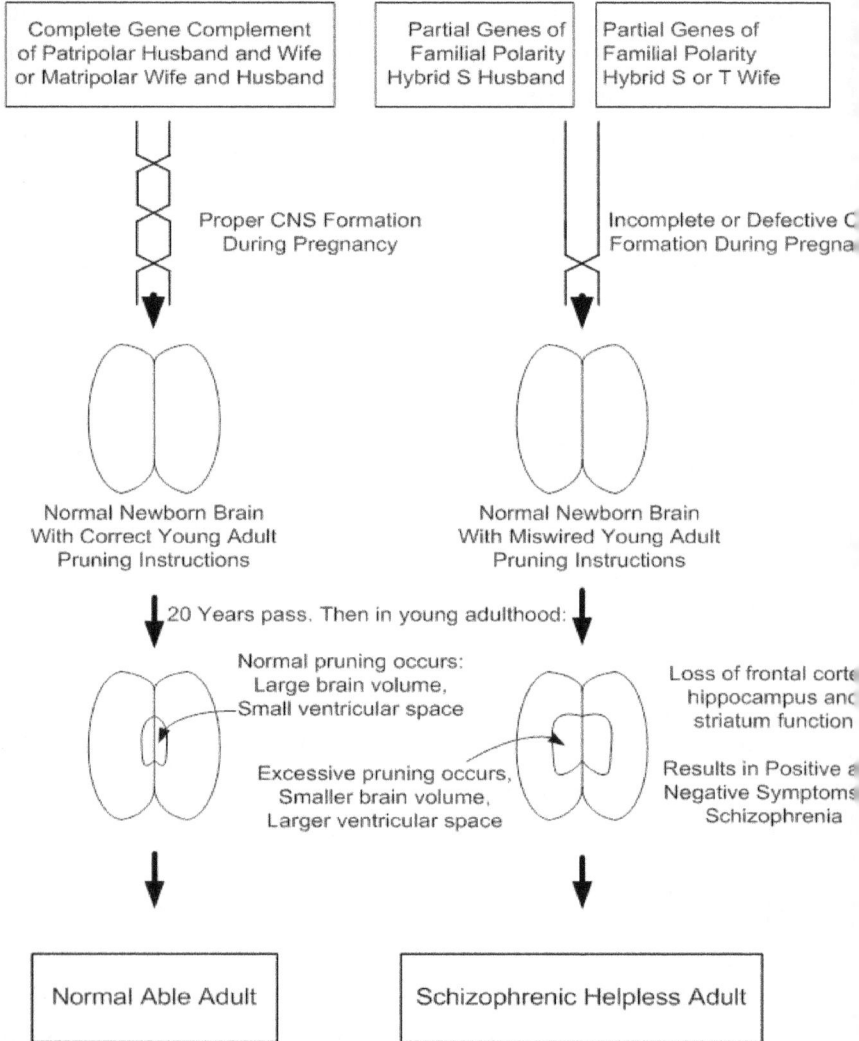

Complete Gene Complement of Patripolar Husband and Wife or Matripolar Wife and Husband	Partial Genes of Familial Polarity Hybrid S Husband	Partial Genes of Familial Polarity Hybrid S or T Wife

Proper CNS Formation During Pregnancy

Incomplete or Defective C Formation During Pregna

Normal Newborn Brain With Correct Young Adult Pruning Instructions

Normal Newborn Brain With Miswired Young Adult Pruning Instructions

20 Years pass. Then in young adulthood:

Normal pruning occurs: Large brain volume, Small ventricular space

Loss of frontal corte hippocampus and striatum function

Excessive pruning occurs, Smaller brain volume, Larger ventricular space

Results in Positive a Negative Symptoms Schizophrenia

Normal Able Adult

Schizophrenic Helpless Adult

striatum and other structures. These losses damage the matching system and short-term memory systems. Both the positive and negative symptoms of schizophrenia can only be the result. Early treatment with neuroleptics may reduce these losses somewhat, but at what cost. These side effects are felt by many to be worse than the illness itself. For a different take, it has been reported that 40% of schizophrenics had been subjected to childhood and sexual abuse.

Figure 17 reviews the nature and extent of familial polarity hybrid pathology for schizophrenia.

The Demons That Drove Mima to a Terrible End.

The following from the Sydney Morning Herald in 2012 illustrates the tragedy of programmed brain digestion in early adulthood:

"To her family, she was Mima. To her friends, she was Joy." When Melissa Joy Dietzel was 18 and left her home in Redlands, California to study at Brigham Young University in Utah, her parents expected wonderful things for her. "We thought she was going to take the world by the tail," her stepfather, Jay Gunther, told Fairfax from his home in Redlands. But yesterday, the family and friends of Ms Dietzel, 22, were told by police and US embassy officials in Australia that her remains had found 30 feet up a tree in Randwick about a month after she had disappeared. It appeared that Ms. Randwick had taken her life.

Three years ago, her older brother Jared killed himself at the age of 24, and Ms Dietzel asked if she could be the one who gave the eulogy at his funeral.

Jared was diagnosed with schizoaffective disorder, a severe form of bipolar disorder, and had two previous suicide attempts, Mr Gunther said. His mother Vickie Gunther's family also struggled with the illness, with both her parents and two sisters taking their lives.

But while Jared wrestled with bipolar illness for many years, Ms Dietzel was the perfect daughter - excelling in her studies, popular at school, and a leader to whom her family and friends looked up to. Ms Dietzel was born on March 19, 1989 to a Mormon family in Redlands, CA. She was sixth in a family of 10 children - with five brothers and four sisters. Her parents, Mrs Gunther and Richard Dietzel, split when she was one and he moved to south Utah.

Ms Dietzel was quiet as a young child, but as she grew up and joined both her school's jazz and marching bands, playing percussion instruments such as the marimba, she became gregarious and had a lot of friends. By the time she went to Brigham Young University, in Utah, Mrs Gunther said. "She worked at the school's counseling center as a student counselor and was always in demand from her bosses and fellow students. She was so good at what she did and so good with people that pretty soon at the office, everything kind of evolved around her. And they would say, 'Melissa, how should we do this?'," Mrs Gunther said. "When she would come home from vacation, even if it was a week or two, her her boss would call her and say, I know we said you could go home, but everything is awful here when you are not here. Nobody is having any fun and no one gets excited about work. We miss you, can you come back?"

Ms Dietzel graduated from Brigham Young with a degree in elementary education in three-and-a-half-years instead of the usual four years and paid most of her university fees herself.

But while Ms Dietzel was self-directed and focused during her university years, things started to change in the months after she returned to her family home in Redlands in 2010.

"She had been a daughter who always accomplished things, but when she returned to stay with us, she didn't have a plan and she wasn't following through. It was not like her," Mr Gunther said. "She was not sleeping, a real common symptom of bipolar disorder ... and she had pressured speech, where she would talk endlessly. It was the first time in our lives [that we had seen her like that]."

Ms Dietzel had brushed off her mother's concerns about her health, telling her that her lack of sleep was just a result of living the life of a university student, and did not see a doctor. "We didn't realize what was wrong," Mr Gunther said. "She was conscious [of what was happening] but she didn't think there was anything to be worried about."

By November 6, 2010, Ms Dietzel was on a plane to Australia on a six-month working visa, and started work as a live-in nanny in Sydney soon after. But her condition worsened. The sleepless nights continued, and she began to hear voices and behave strangely, Mr Gunther said. Her employer let her go after a short while, thinking she might return to the United States. Ms Dietzel had a return ticket and told him she had to go home anyway as the voices had told her that her parents were dying.

Later, when her employer her stepfather he had wanted Ms Dietzel to undergo a psychiatric evaluation, and that she appeared to be having symptoms of schizophrenia or bipolar disorder.

The last time Mr Gunther heard from her was on November 30, and despite the family's attempts to contact her through email, Facebook and by phone, Ms Dietzel failed to respond. They filed a missing person's report with Australian police. They hoped she might have wanted some personal space and believed she would have had to return to the United States in May when her visa expired. But last week, as the family prepared for the wedding of her sister, Amber, US police contacted the family and asked for her dental records.

Ms Dietzel was identified through the records and her family was formally notified yesterday. Her mother said the family plans to cremate her and have her ashes sent back to Redlands with her brother Nicholas, a sculptor who was making her urn. "He's very artistic and he makes custom urns for people. He said, 'I would be honored to make Melissa's.' And that's what he's going to do."

A similar scenario as this one for schizophrenia could be developed to account for the delayed defects occurring in obsessive-compulsive disorder and bipolar disorder (17). Correspondingly, the mechanism producing dyslexia, could be extended to account for autism, epilepsy, mental retardation, Tourette's syndrome, and ADHD.

CHAPTER 9. ORIGIN OF GOTHISM AND THE DARK SIDE

All Familial Polarity Hybrids Have a Basis for Self Loathing: Some More than Others

Their self-hatred comes from their inability to meet societal or survival norms naturally, and from the rejection that such failure brings. It comes from the dawning recognition that through no fault of their own, being normal will be forever impossible for them. Self-loathing comes from having futile, non-functional, and misdirected abilities, orientations and tastes. All the familial polarity hybrids of this book must deal with their self-hate.

Probably the family polarity hybrid that feels the smallest but still significant self-loathing is the partially compensated dyslexic. They often are able by their extra intense effort to rationalize that they are normal. Yet, there is self-resentment that they must work continually just to keep up with others who have plenty of free time to enjoy life.

For the failing dyslexic, this self-disgust and feeling of inferiority must reach much higher levels. Not being able to read well and too slow of thought to be competitive for good jobs, they fall into lower economic levels. The exceptional self-made dyslexic genius is a little happier. But he still hates himself for his endless mental blocks, forgetfulness, and lack of glibness.

The trans-heterosexual feels significant amount of self-hatred coming from being sufficiently different from normal as to attract negative attention at primary school. They often become victims of early teasing, ridicule, and bullying. Later, when a hetero-male with a feminine identity compares himself to a masculine male, it is obvious that something is missing. He is at a loss to compete and ends up appearing wimpish in contrast. That has to hurt. He must do something artistic or intellectual to compensate, but he still hates himself.

Similarly, when a hetero-female with a masculine identity is compared with a feminine woman, their husbands can only complain of their harshness and their relative asexuality. Try as they might, they just do not feel feminine. The difference is obvious, and at some level, they hate themselves for not being a real woman.

However, when it comes to the cis-homosexual gays and lesbians with normal sexual identities, they must confront the huge shock and rejection that comes from society when they act on their normal but now unnatural drives to mate with members of the same sex. This causes them especially to loath themselves. They feel deep resentment for not being naturally attracted to the opposite sex.

Transvestite hybrids can only hate themselves for the contempt poured upon them by normal people when they are true to their cross-wired feelings. Cross-dressing and lovingly embracing others of the same sex seems so bizarre to normal people as to be dangerous to the exposed hapless hybrid. Rejection, ridicule and bullying in school are inevitable.

In a different way, self-loathing comes even worse for the schizophrenic. He or she has been wrenched from a familiar normal 20 years of life by an aberrant last phase of brain development. His or her brain, being attacked by inappropriate cell suicide commands, has been shrunken dreadfully in major areas. They are left with little memory and a contextual confusion so great that they often cannot tell what is real and what is not.

The self-loathing is particularly great in younger patients while they still may have an insight into the serious permanent effects that the illness will have on their lives. The schizophrenic hates himself for his permanent incapacitation and loss of access to the previous joys of a normal life.

Lastly, none of the familial polarity hybrids commonly cause as much harm to others except for the hated pedophiles. In their uncontrollable lust, each of them scars hundreds, sometimes thousands of innocent defenseless children. Their naturally strong but misdirected reproductive drive makes life a living hell. They suffer from deprivation if they don't act on it, and harm others and becoming criminals if they do. Talk about a reason to hate oneself.

Reactions to Self-Loathing: Drug Abuse, Depression, Self-Mutilation, Suicide

Self-hatred is not a natural state of mind. It can only come after many failures to reach normal goals. These endless, inevitable failures make it clear that for the hybrid, normal behavior for them is impossible. Their impulses cannot be trusted to be their guide, as it can in wild types. This leads to despair, a state of mind where the reward neurotransmitter, dopamine, becomes depleted. Months of despair and self-loathing lead to depression, a mental state where reward is lost and only pain, failure, hopelessness and death remain. Suicide rates are very high among the depressed. It can safely be said that suicide rates among familial polarity hybrids in general are elevated over those of normal non-familial polarity hybrids.

In many cases, self-loathing and depression can lead to hostile attacks against one's own body. Research shows that as many as one of every dozen teens harm themselves by deliberately cutting and burning themselves and, in some cases, by suicidal acts as they progress from puberty into young adulthood. Schizophrenics have been known to cut off their own penises in a desperate, misguided effort to manage their still present reproductive drive.

Those who cut, burn, or otherwise deliberately hurt themselves are more likely to be seriously depressed or anxious, and to report smoking, drinking, or drug abuse (23). This behavior can artificially elevate the missing dopamine to give temporary relief from dysphoria. Unfortunately for them, tolerance and withdrawal anguish soon follow.

Shifting the Blame by Turning to the Dark Side

Once they discover by prayer and meditation that religion cannot cure their cross-wiring anomalies, they often reject any creator God who would permit such deviations of nature to occur. To them, the greater power in

have long existed. They begin to relish their deviations, calling themselves "The Living Dead". They begin to understand their situation.

For some of them, evil overpowers good. They view that acting out their unnatural cross-wired feelings is their highest goal, regardless of its inherent harm to life. This turns the psychopaths among them toward fangs, blood lust, corpse mongering, human sacrifice, serial killing, and cannibalism.

Enter the Goth Subculture:
The Goths were a barbaric East Germanic tribe of Scandinavian origin whose two branches, the Visigoths and the Ostrogoths, played an important role in the fall of the Roman Empire and the emergence of Medieval Europe.

The term gothic, as in "Gothic architecture", was an epithet by Italian architects referring to a so called barbaric building style that flourished in the medieval period. The Italians called this later magnificent style, crude or Gothic in comparison to classical Greek and Roman architecture.

Gothism is a contemporary subculture found in many countries. It began in England during the early 1980s in the gothic rock scene, an offshoot of the post-punk genre. The goth subculture has survived much longer than others of the same era and has continued to diversify. Its imagery and cultural traits indicate influences from the 19th century Gothic literature, along with, horror films, and to a lesser extent the bondage, discipline, sadism and masochism (BDSM) culture.

Gothic literature, sometimes referred to as Gothic horror, is a kind of fiction that combines elements of both horror and romance. Gothicism's origin is attributed to English author Horace Walpole, with his 1764 novel The Castle of Otranto, subtitled "A Gothic Story". The effect of Gothic fiction feeds on a pleasing sort of terror, an extension of Romantic literary pleasures that were relatively new at the time of Walpole's novel. Melodrama and parody (including self-parody) were other long-standing features of the Gothic initiated by Walpole.

Archetypes in the Gothic novel included the Virginal Maiden, Older Foolish Woman, Hero, Tyrant, Stupid Servant, Clowns, Ruffians, and Clergy (always weak, usually evil).

The proper setting of the Gothic novel is essential. The plot is usually set in a rundown castle or some other usually religious edifice. It is to be assumed that this building has secrets of its own. This gloomy and frightening scenery not only evokes the atmosphere of horror and dread, but also portrays the deterioration of the world to hopelessness and evil. Without the decrepit backdrop within which to initiate the events, the Gothic novel would not exist.

Horror films seek to elicit a negative emotional reaction from viewers by playing on the audience's most primal fears. The term "horror movie" first appears in the writings of critics and film industry commentators in response to the release of Universal Studio's *Dracula* (1931) and *Frankenstein* (1931). Horror films deal with the viewer's nightmares, hidden worst fears, revulsions and terror of the unknown. Plots written within the horror genre often involve the intrusion of an evil force, event, or personage into the everyday world. Themes or elements often prevalent in typical horror films include ghosts, torture, gore, werewolves, ancient curses, satanism, demons, vicious animals, vampires, cannibals, haunted houses, zombies and serial killers.

Bondage, Discipline, Sadism, and Masochism among the Goths is an erotic preference and a form of sexual expression involving the consensual use of restraint, intense, often painful sensory stimulation, and fantasy power role-play. BDSM includes a wide spectrum of activities, forms of interpersonal relationships, and distinct subcultures.

The participants usually taking on complementary, but unequal roles characterize activities and relationships within a BDSM context. Thus, the idea of consent of both the partners becomes essential. Typically, participants who are active – applying the painful activity or exercising control over others – are known as tops or dominants. Those participants who are recipients of these activities, or who are controlled by their partners, are typically known as bottoms or submissives. "Love is pain, so please hurt me more," is a statement a bottom might make

According to Wikipedia, Goths, in terms of their membership in the subculture, are usually not supportive of violence, but are tolerant of alternative lifestyles that incorporate themes such as BDSM-always involving consent. Violence and hate do not form elements of gothic ideology; rather, the ideology is formed in part by identification, and somewhat of a passion or interest in death and darkness, grief over societal and personal evils that mainstream culture wishes to ignore or forget. These are the prevalent themes in gothic music.

Observation #32: Most Goths are Familial Polarity Hybrids.
Why Gothism Repels Normals and Attracts Hybrids

Who in their right mind would cut and mutilate their bodies, suffer from depression, or from the abuse of drugs? Who would mix pain and degradation with sex? Who would worship evil and be drawn toward darkness, corpses, blood, and gore? Who would become cynical and alienated to mainstream society, have an ambivalent or tragic sexuality and be gripped by morbid motifs?

These attitudes are not the natural choice of wild type normal Cis-Hets. In fact, they are quite the opposite. To normals, their bodies, which they can trust, are holy temples that fill them with pleasure. They mutually draw partners with love to create sacred families within which to raise normal children. They migrate toward life and the good.

Instead, Gothism appeals to the hopeless, helpless, the out of control, the failures, the weak, the repulsive, the ostracized among us. That is Gothism appeals to familial polarity hybrids for the reasons outlined in the beginning of this chapter.

Teens are in a Psychosocial High-Risk Culture

The Goth culture attracts hybrid teens, who are depressed, feel persecuted, have a distrust of society, or have suffered past abuse. They surround themselves with people, music, Web sites, and activities that foster angry or depressed feelings. They have a higher prevalence of depression, self-harm, suicide, and violence than non-Goth teens.

Mass Murder Revenge

Although Goths disavow the killers among us as true Goths, the overriding theme and context of hybrids suffering bullying and abuse is a hopeless anger that can only be escaped by death. Many hybrids, even if not doctrinaire Goths, are brilliant in their own way. If emboldened by partnership and ridiculed sufficiently, they have been known to strike back. They have decided to kill as many their normal tormentors and others as they can in the process of escaping by death themselves. The Columbine High School Massacre is the classic case of this kind. Of course, it is not the fault of wild types that they are normal. Nor should they have to pay.

Gothic Growth

On the other hand, for the lonely hybrid to discover that he or she is not alone must have powerful benefits. Perhaps the first wave of Familiar Polarity Hybrids who grew large and successful enough to become a counter culture was the Beatnicks of the Beat (down and out) Generation in the 1950s.

The Hippie revolution of the 1960s reversed this negativity with "Flower Power". However, after the governmental criminalization of psychedelic drugs, the negativity returned as antiestablishment Punk Rock in the 1970s. *The Rocky Horror Show* was produced during this time. The Rockers included gays, lesbians, and trans-sexuals whose goal was to make a statement that normal people would find repulsive, obscene, and horrifying. They thrived upon negative attention from their unorthodox dress, hairstyle, makeup, and rebellious, crude music. They rejected Elvis, the Beatles, and the Rolling Stones, substituting it with their "No Future" nihilism.

In the 1980s, Goth appeared as an offshoot of the post-punk genre. This dark theme has become amplified and numbers of followers have greatly increased. The shock value created by Goth-related entertainers has made them enormously wealthy superstars known and beloved by all Hybrids. They include Madonna with her sacrilege; the gifted pedophile Michael Jackson; with his large children's play park; the self-cutter, bisexual, Johnny Depp; and the shocking Lady Gaga, with her raw meat attire and her boney shoulder and skull "implants".

Each of these entertainers is or was brilliant and talented. They have made a fortune appealing to the self-loathing and despair associated with their lives of familial polarity hybrids. The size of their adulating audiences gives some idea of the number of hybrids existing, if one could subtract the number of normals paying to see these provocative side-shows.

So, does this popularization of the Dark Side make it easier for FP hybrids to exist? Can such a death embracing anti-life movement enable them to shift blame for their undeserved deviance to fate and thus live a more fulfilling, stress free life?

Early schooling is a key time where the lives of FP hybrids could greatly be improved. They could be identified and informed of the true nature of their cross-wiring defect, and told that it is a common occurrence. They could be placed in groups of similar others, and educated appropriately. Membership in such a family could give resilience to external conflict and greatly improve their self- images.

Past Overreactions to the Proto Goths and Other Hybrids

Let us not forget that RM Adolph Hitler noted that "pollution" of so-called Arian Blood by interbreeding of the German master race with Slavs and Jews was occurring in his day. According to him, these mixed marriages produced inferior undesirable offspring, including homosexuals, the mentally retarded (dyslexics), and the mentally ill (schizophrenics). The elimination of these populations was partly the basis of his genocide programs. Later, Serbian LM Slobodan Milosevic made similar observations, which in part led to his ethnic cleansing programs against the Albanians and Croats.

Both were correct in their observations of defects in hybrids. But, they totally wrong in their conclusions that the polluting race was not human, inferior, and deserving of annihilation. Actually, it was the hybrids that were "inferior", not the Jews, Slavs, Albanians or Croats, who themselves are comparable to their opponents. For example, Jews are highly represented among Nobel Prize winners. To use systematic genocide against the other race as their "cleansing method" was totally misguided and repugnant, leaving a huge scar on civilization. However, killing dyslexics, homosexuals, pedophiles Goths, and schizophrenics is obviously an impractical answer as well, since they appear to constitute at least half of the earth's population.

The ultimate utopian solution will come first from the use of DNA sequencing to distinguish the two familial polarity species, along with their many hybrids. Then, marital counseling can be used avoid harmful combinations. Later, genetic engineering can be used to correct inappropriate cross-wirings. In the mean time, let us, for the first time become clear about the existence of familiar polarity, its consequences, its hybrids, and their characteristics

.

CHAPTER 10. FAMILIAL POLARITY REWRITES HISTORY

We will put cross-breeding and hybridization aside in this chapter, as it is only a part of a much larger story. The discovery of the two human species of familial polarity has opened a new window, not only for clarifying human origins and history, but also for understanding current global conflict.

As illustrated in **Figure 19**, post ice-age migrations out of Africa appear to have created foundational human populations (3, 26, 50) of alternating familial polarity striations across the eastern hemisphere. In this data-consistent, but speculative scene, apparently the highly intelligent patripolar orangutans were the first out of Africa. They traveled eastward along the coast into the sunrise to arrive in Northern China and the Yellow River Valley almost 2 million years ago as the Peking Man.

To account for how thay got so far north by following the coastline, it appears that they followed the then southern coastline of Asia before the Indian subcontinent had docked with Asia. After that docking, another wave of orangutan-Shivapithicus stock continued out of Africa to the east, but the second time followed the newly available southern coastline of India, this time arriving in Indonesia where they now reside. It is interesting that today orangutans are totally extinct in Africa with only fossils remaining. It is also entertaining to speculate that the

Figure 19: Out of Africa Migrations to Locations in the Eastern Hemisphere Resulted in Striations of Alternating Familial Polarity

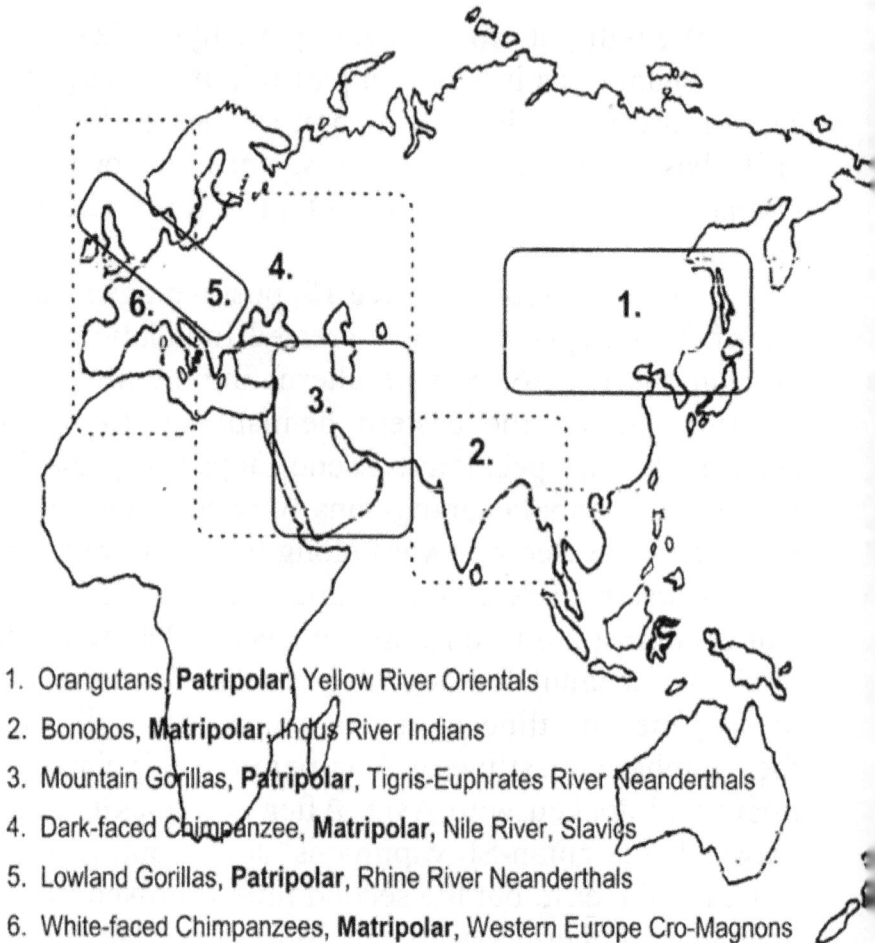

1. Orangutans, **Patripolar**, Yellow River Orientals
2. Bonobos, **Matripolar**, Indus River Indians
3. Mountain Gorillas, **Patripolar**, Tigris-Euphrates River Neanderthals
4. Dark-faced Chimpanzee, **Matripolar,** Nile River, Slavics
5. Lowland Gorillas, **Patripolar**, Rhine River Neanderthals
6. White-faced Chimpanzees, **Matripolar**, Western Europe Cro-Magnons

distinctive orange color of the orangutans could have been be the origin of the yellow, orange, and red pigmentation in the oriental races.

Second, the black, straight-haired matripolar bonobos were apparently the next to break out of Africa. Also moving eastward, they failed to migrate much beyond India, perhaps because of population build up from earlier more eastern migrant orangutan stocks. Thus, we now find the Dravidian black, straight haired human stock of the Indus River valley and southern India.

Third, the Mountain Gorillas migrated into the Tigris Euphrates River Valley to become part of the Neanderthals.

Fourth, hominid derivatives from the matripolar Dark-Faced Chimpanzees then took over the Nile River Valley, additionally migrating into Eastern Europe as the Slavic races.

Figure 20 shows remains of the Dminasi Man, the first Caucasian. These bones were found between the Caspian and Black Seas of the Caucasus near the ancient city of Tbilisi, Georgia, far from Africa. His era is dated at 1.7 million years ago. His cranial capacity (**Figure 20**) was 700 cm^2, only half of ours. Various stone flakes scrapers and chopping tools were found with these remains.

Fifth, Lowland Gorilla-derived hominid stock migrated north to become the patripolar Neanderthals of Rhine River and Western Europe.

Figure 20. The Dmanisi Man with tools, the first Caucasian, dated at 1.7 million years bp

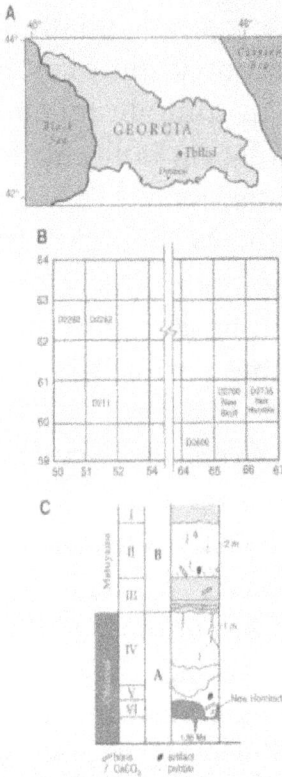

The skull is in remarkably fine condition (Fig. 2). The maxillae are slightly damaged anteriorly, the zygomatic arches are broken, and both mastoid processes are heavily abraded. There is damage also to the orbital

be a female. However, the upper canines carry large crowns and massive roots, and their size counsels caution in assessing sex.

In its principal vault dimensions, D2700 is smaller than D2280 and the specimens attrib

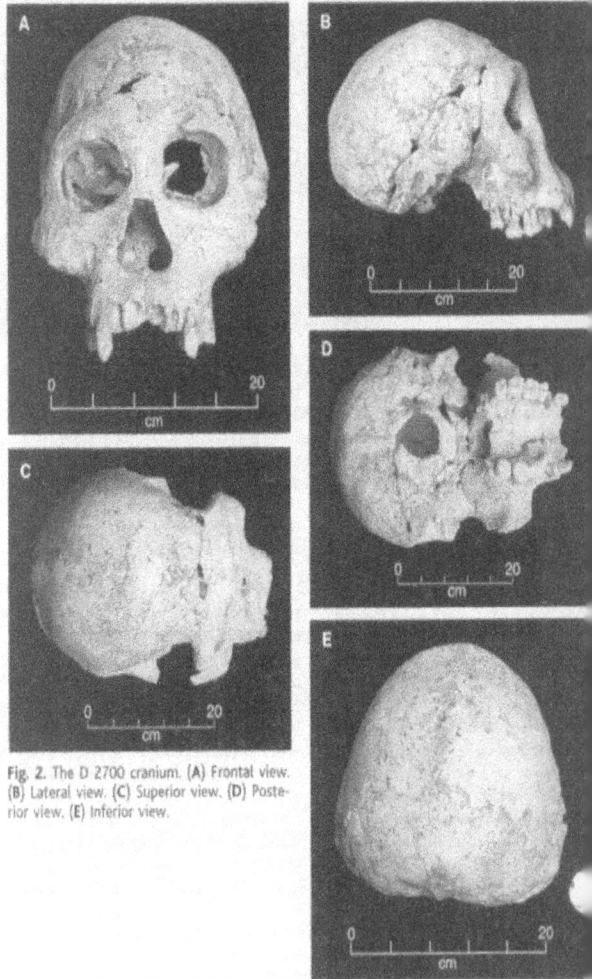

Fig. 1. (A) Location map of Dmanisi site. (B) The locations of hominid fossils (excavation units are 1-m squares). (C) General stratigraphic profile, modified after Gabunia et al. (5, 6). The basalt and the immediately overlying volcaniclastics (stratum A) exhibit normal polarity and are correlated with the terminus of the Olduvai Subchron. Slightly higher in the section, above a minor disconformity and below a strongly developed soil, Unit B deposits, which also contain artifacts, faunas and human fossils, all exhibit reversed polarity and are correlated with the Matuyama. Even the least stable minerals, such as olivine, in the basalt and the fossil-bearing sediments show only minor weathering, which is compatible with the incipient pedogenic properties of the sediments.

Fig. 2. The D 2700 cranium. (A) Frontal view. (B) Lateral view. (C) Superior view. (D) Posterior view. (E) Inferior view.

174

Sixth, the White Faced Chimpanzees evolved into the blue-eyed blonde matripolar Berbers who migrated repeatedly across Gibraltar to become the Western Mediterraneans and Scandinavians. Thus, we have at least six alternating familial polarity striations of hominid stock populations established anciently across the Eastern Hemisphere. This has further consequences to be described later.

Prehistory of Familial Polarity

Anciently, small but growing populations of humans were distributed in such a way that these patripolar and matripolar stock striations were generally isolated from each other. One married someone from one's own village, such as the girl next door, thus keeping both lineages pure. However, things changed as populations increased and overlapped into cities. Then one might marry someone from work instead from one's own tribe, often someone of opposite familial polarity. In addition, marauding patripolar barbarians coming from the east on horseback endlessly raped and pillaged matripolar farmlands. This resulted in the production of large numbers of hybrid offspring, some of which were noted even by Greek and Roman times to be different from the normal human stock in terms of sexual orientation.

Two prescient independent yet parallel syntheses of the prehistory of human conflict were written in the absence of knowledge of dyadic evolution and its consequent familial polarity. In *The Chalice and the Blade* (33) and *Sacred Pleasure* (34), Riane Eisler clearly distinguished early female dominant "Gylanic" cultures, which were obviously matripolar in lineage, from the later destructive male dominant "Dominator" cultures, which clearly were patripolar. In *Saharasia*, James DeMeo also traced the rise of desert "Armored Patrism", around 4000 BC (26) with the domestication of the horse (30). Horsemen then began to prey upon the low-violence forest "Unarmored Matrism" cultures of the Old World. With the here-added focus of Familial Polarity, each of these well-documented theses can be seen to illustrate and clarify further the existence of matripolar and patripolar cultures and their inevitable conflicts (75, 152).

In **Table 19**, a brief summary of earth history since the last Ice Age 14,000 years ago, places these familial polarity conflicts in perspective.

Familial Polarity and the Origins of Judaism:

Biblical Abraham was a successful patripolar herdsman from the city of Ur in the Tigris and Euphrates river valley of Iraq. Taking advantage of a temporary wet period, he migrated with his flocks and herds south into Palestine. With him, he brought his patriarchal religion, based upon Laws of Hammurabi, and the Gilgamesh Epic. In Palestine, Joseph, a favorite son of Jacob, Abraham's grandson, was sold to slave traders traveling to Egypt.

There, Joseph was sold on the slave block to an upper class Egyptian matripolar RF woman to be her household assistant, and not unexpectedly, her stud. Sexually inexperienced RM Joseph from the patripolar "virginity until marriage" culture spurned her amorous

Table 19, The Post Ice Age History of the Current Era.

12,000 BC: Ice recedes, Forests regrow to cover the earth, Present era begins
12,000 BC: Impossible monumental stone megaliths are built by the Unknowns
10,000 BC: Plant abundance leads to unrestricted human population growth and spread
 9,900 BC: Oldest continuous tree ring record reaches back to this point in time
 9,000 BC: Clovis arrowhead points are left behind in the North America
 8,500 BC: Decline of big game in Northern Hemisphere due to overhunting
 8,000 BC: Game animals are domesticated: cattle in Africa; goats in Iraq;Herding grows
 8,000 BC: Crops are domesticated in Near East: Wheat, barley, lentils: Forests cleared.
 7,000 BC: Matripolaric Cooperator civilizations. Unwalled cities, rich in art, technology
 6,000 BC: Desertification begins: exploited high pop. areas: Sahara, Arabia, Gobi fail.
 5,000 BC: Starvation and battles for limited resources begin. Patries adapt to badlands.
 4,500 BC: Patripolar violent reproductive strategy & greater size. defeats Matri-males.
 4,000 BC: Horse domesticat. by Patries on Ukraine steppes. Nomadic conquests begin.
 3,500 BC: Mounted patripolars with metal weapons cross Old World, "Live off the land",
      raping, killing, burning, raiding, 'til agriculture was impossible, except in spots.
 3,300 BC: Otzi got arrow in neck in Alps:Tattoo acupuncture points on back,copper axe.
 3,000 BC: Patries parasitize cultures. Keep themselves separate,install Caste Systems.
      Forests cut for ships & to smelt metal weapons. Moses overgrazes. Desertification.
 2,500 BC: Horseback child rearing creates psychoses. Female genital mutilation, Ritual
      Widow murder, Cross-polar hybrids are common, Homosexuality rampant.
 1,500 BC: Patripolar Greek Golden Age begins.
  800 BC: Matripolar Roman males suppress females, dominate w. patri- mercenaries.
   33 BC: Rxm dyslexic Jesus is Born, and the roots of Christianity are laid.
  313 AD: Lm Constantine supports Christianity as the religion of the Roman Empire
  622 AD: Rm Mohammed founds Islam.
 1000 AD: Dark Ages of religious domination and cultural stagnation.
 1517 AD: Rm Luther, the Protestant Reformation, and European Renaissance begins.
 1859 AD: Rxm dyslexic Darwin publishes his Origin of the Species.
 1915 AD: Rxm dyslexic Einstein develops the theory of general relativity.
 1945 AD: Atom bombs were dropped on Japan.
 1953 AD: DNA structure is published.
 2003 AD: Human DNA Genome is sequenced.
 2015 AD: RxM dyslexic Morton recognizes two trees of life and the two human species.

advances. Furious, she had him thrown in jail with a five-year sentence to force him to grow up.

While in an Egyptian prison, Joseph repeatedly demonstrated his right-brain big-picture facility for the interpretation of dreams. This greatly impressed his cellmate, one of Pharaoh's wine tasters, temporarily jailed due to palace intrigue. Later reinstated, the wine taster became aware of dreams troubling the Pharaoh himself. At the wine taster's recommendation, Joseph obtained the opportunity again to apply his skill with symbolism and metaphor. He interpreted Pharaoh's dreams in such a satisfying manner that the monarch ultimately made him Prime Minister. Joseph was made responsible for bringing about his proposed solution to the king's dilemma. Thus, he supervised the building of an extensive grain silo system to store the abundance arising from the "fat years" of temporarily abundant rainfall.

A few years later, the wet period ended and the "lean years" came. The drought decimated Abraham's flocks and herds still in the Levant. To ward off starvation, Abraham's sons traveled to Egypt, by then known to be an abundant source of food. There, Joe gets the last laugh on his brothers. Later, with his help, his father's family immigrates to Egypt for eight generations. (How many of us today know the names of our ancestors eight generations earlier?) There, during those roughly 300 years, they freely interbreed as slaves, and thus became genetically transformed into matripolars, unlike their still-patripolar Arabian cousins who remained behind.

It is of interest that, in addition to the Egyptians of Joseph's days being matripolar, the Persian Iranians of the 50 years of Israel's Babylon Captivity were also matripolar. Regardless, it is clear that most Jewish couples today are LMs-RFs, exhibiting the famous "Jewish Princess" and "Jewish Mother" female dominant syndromes, while most Rabbis are detail-oriented, matripolar

LMs, paradoxically still clinging to the details of their ancient patriarchal religious traditions. These opposing elements of familial polarity have a direct bearing on the present seemingly impossible relations between the matripolar Jews and the patripolar Arabs, and importantly between the matripolar Iranian (Persian) Shiites and the patripolar Arabian Sunnis. These family pairs of opposite familial polarities are as miscible as oil and water.

Familial Polarity and the Origins of Christianity:

Several lines of evidence suggest that Jesus of Nazareth was a RM, unlike most of his LM contemporaries. He was different. Jesus had cosmology skills that modern RM sons have early, but LMs gain only later. These, he demonstrated at age of seven to the priests in the temple.

His RM nature also showed itself through his emphasis on need for the love of a divine *father*, such as in his "Lord's Prayer". Neither unwed mother RF Mary, nor her later husband, LM Joseph, could have substituted for the genetic need of Jesus for a dominant RM role model father (Chapter 10). Neither would RF Mary have been able to give him the unconditional love RMs need from their LF mothers. Thus, he distanced himself from his family, including Mary and his LM half-brothers. If we were to deny Mary's "unique in all history" claim of Virgin Birth, her RM son could only have come from a hidden intimacy with a patripolar male, such as a Roman soldier or Arab merchant. As a result, it is likely that Jesus may have actually been an RxM with unnoticed developmental dyslexia like Einstein and Darwin.

Jesus' possible RM nature is supported further by his creation and extensive use of politically correct seeming, right brain-derived metaphorical parables with hidden meanings that broke with tradition: "He who hath an ear, let him hear". In addition, what would not have been expected from a LM, was his propensity to meditate, his willingness to cry, his emphasis on emotionality in the Sermon on the Mount, and his violent driving of the moneychangers out of the temple.

***Provocative Speculation:* One of Several Alternative Interpretations of the Crucifixion, Resurrection, and Translation of Christ**

Part A: Voodoo, Zombies, and Puffer Fish

Zombification is part of the African Voodoo witchcraft tradition found among former slaves in Haiti. The victim is cursed, treated with a secret potion, falls into a deep coma, is pronounced dead (often by a western physician), and buried. In the *Night of the Living Dead*, he or she is dug up within 6 hours by the witch doctor, called forth, resurrected, drugged with scopolamine, and led away as a mindless zombie slave for abuse by distant sugarcane grower landowners.

Tetrodotoxin is a sodium ion channel blocker found in puffer fish skin and viscera. Since sodium ion is essential for nerve transmission, poisoning symptoms range from mild euphoria (as in Japanese Fugu cuisine), to prolonged coma, to death. Several revivals "from the dead" have occurred in modern morgues after puffer fish poisonings. Woodard at Harvard University determined the structure of Tetrodotoxin. Wade Davis (Movie: *Night of the Living Dead*) obtained zombie potions from Haitian witches and took them in to Harvard for analysis where significant amounts of Tetrodotoxin was found in them, **Figure 21**.

Part B: Possible Tetrodotoxin Poisoning of Jesus on the Cross

Late Friday, the Romans nailed Jesus to a cross between two others on the hill called Golgotha. After hours of anguish, he cried out, "I thirst!" On a pole, his followers raised to him a sponge dipped in fish gall (which can contain tetrodotoxin and Roman soldiers commonly used in crucifixions), which according to John, he took. Not long after he drank this, he said "My God! Why hath Thou forsaken me?" Then, he said "Into thy hands I commit my spirit". "It is finished!" and he "died" (that is, he went into a two day tetrodotoxin coma, awakening early Sunday morning).

As the soldiers wanted the victims off the hill by Friday night for the feast of the next day, they broke the legs of the still conscious thieves on either side to hasten their death. Jesus, in a death-like coma, was only stabbed in the side before being taken down. Joseph of Aramathea and the "wise man", Nicodemus, asked Pilate for Jesus's body. It was placed in a private stone crypt, covered by a stone door and "guarded" by soldiers (107b).

Figure 21: Scientific Zombification

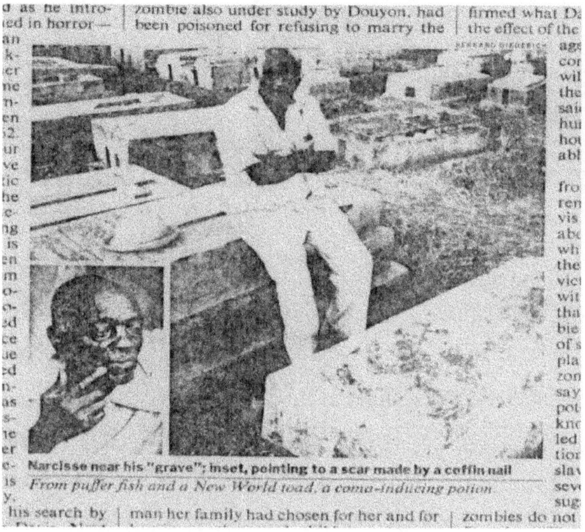
zombie also under study by Douyon, had been poisoned for refusing to marry the
Narcisse near his "grave"; inset, pointing to a scar made by a coffin nail
From puffer fish and a New World toad, a coma-inducing potion
man her family had chosen for her and for
zombies do not

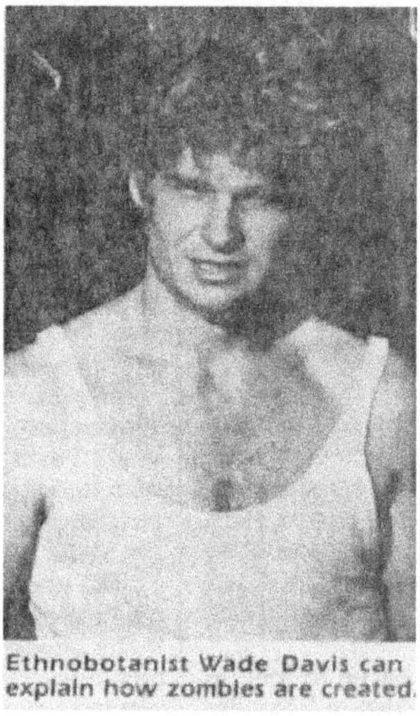
Ethnobotanist Wade Davis can explain how zombies are created.

Time Magazine Oct 7, 1983

Movie: The Serpent and the Rainbow, 1983

Apparently, Jesus recovered consciousness from the tetrodotoxin coma early Sunday morning. As is written in the New Testament, he was seen by many of his followers several times over the next few days. During that time, Jesus asked them to meet him at a previously mentioned mountain site in Galilee. When they all met on the mountain, he exhorted them to: "die to the flesh, be reborn of the spirit, and to go forth and teach all nations in my name." Then, he lifted his hands and blessed them. While blessing, he parted from them into the clouds (i.e., he walked up the hill and disappeared into the cloud base of the local marine inversion layer). They are said to have returned to the Temple in Jerusalem, praising God continually.

Later, Jesus apparently migrated to the south of France (12c), where he returned to a profession of changing water into wine. There he was known as God's child (i.e., Rothschild: *humor)* Thus, did Christianity originate from the life, "death", "resurrection", and "transfiguration" of RxM, Jesus.

By the time of the Aryan Controversy (over the proposed divinity of Jesus) and of the adaptation of Catholic Christianity as the official state church of matripolar Rome, in the fourth century by Constantine, Mary had become "The Mother of God".

After the more than a millennium of "Dark Ages" following, the 16th century Protestant Reformation began, initiating the modern era. It started in Germany with Martin Luther, a patripolar Catholic priest. Among other things Luther protested was the demasculinization of Christianity and the overemphasis of the feminine in the Mariolatry that then existed. His protest, along with that of others, led to a Reformation of Christian church, and ultimately to the many of the Protestant denominations of today. Protestantism has had its strongest appeal in patripolar population centers, such as in Germany, Switzerland, Scotland, Northern Ireland, and now in within the Republican US "Bible Belt" (84).

Familial Polarity and Its Impact on Modern Forms of Government   As with any new paradigm, a restructuring of the fabric of knowledge occurs with consequent waves and ripples extending from its epicenter. Examples of such exist, not only in the patripolar Protestant Reformation, but also in the patripolar founding and development of the US. This places a new perspective upon the attempts of charismatic patripolar Puritan, Pilgrim, and Quaker males to escape the tyranny of the "moral majority" present in the established matripolar ethnic blocks of the Old World. That is, they wished to enjoy the freedom of belief and worship, which is never permitted under the religious legalism, censure, and oppression imposed by matripolar males wherever they become unified by their hierarchal territorial dogma.

The American Revolution elaborated this freedom theme further under the leadership of such patripolar males as Benjamin Franklin and Thomas Jefferson. In the Civil War, the patripolar Confederates of the white south who wished secede in order to use slaves in a form of economic competition in spite of the wishes of the national majority, again repeated this drive for separation and freedom from external control.

More recently, the founding of the LDS (Mormon) and other patripolar Christian denominations dramatized this patripolar separation-isolation theme. For example, a number of recent US social rebels came from patripolar Seventh-day Adventist, conscientious-objector roots. These include Malcom X, David Koresh, and Lee Malvo. The Mormon-derived polygamists should not be overlooked, either.

Yet, these thematic sub-elements pale in the face of patripolar Islamic history and possible "Battle of Armageddon" familial polarity interpretations of current Middle East Conflict. As a generalization then, it would appear that religion and culture have followed the biology of familial polarity, not vice versa. That is, it is here asserted that all the founders of the world's religions were patripolar males as outlined in **Table 20**.

Familial Polarity and the Logic of National Governance Styles: Autocracy vs. Democracy

Fundamental political differences between the naturally different power structure orientations of the two polarities have been a major obstacle to the achievement of a stabile mixed polarity society. In the past, patriarchal Haremic males not only settled their reproductive rights by physical combat, but also their political leadership. Anciently, the winner of individual combat and consequently the harem leader regularly demanded acts of physical submission from each formerly excluded male (from primate bachelor camps) before allowing him to join his family group. In some modern primates, a submissive genuflection of the subordinate before the alpha male's erect penis is required (79).

Anciently, as a not too far-fetched suggestion, this essential loyalty step brought the benefits of added male partnership to the harem while securely protecting the leader's breeding rights by the establishment of an early form of the death penalty: "Touch my women and I'll kill you!" Later, acts of obeisance, that is, pledges of absolute obedience were demanded of all male followers.

These then became oaths of allegiance, whereby

Table 20: All the World Religions were Founded by Charismatic Patripolar Males:

Judaism:	_Abraham:_ patripolar – Patriarchal, Jews become matripolar after 300 yrs (eight generations) in Egypt
Christianity:	_Jesus;_ patripolar, Early church and Protestantism. Later, Catholicism amalgamated with Roman worship of sun and feminine to become more matripolar: Mariolatry
Islam:	_Mohammed:_ patripolar, male dominance in home and government. Persian Shiites are matripolar and often in conflict with patripolar Abrabian Sunis.
Confucianism:	_Confucius:_ patripolar, male dominant in home and government, Northern Orient
Taoism:	_Lao-Tse:_ patripolar (North China) and matripolar (South China)
Buddhism:	_Gothama:_ patripolar and matripolar Segments: Theravada vs. Miyahana Buddhism
Hinduism:	_Unknown patripolar Arian_: later Dravidian matripolar engulfment
Jainism:	_Vardhamana:_ patripolar : no killing, lying, stealing, adultery, or greed
Sikhism:	_Nanak:_ patripolar

the follower would swear literally upon the loss of his testicles if he were to disobey the orders of the leader, and that his "testi"mony was true. Such laws and "testi"ments were followed implicitly as long as the leader was in power in the *fatherland*. The leader commonly administered spontaneous "tests" of submission to his followers.

The current existence among educated and intelligent modern Haremic males of a powerful underlying dominance psychology is abundantly demonstrated by the extraordinarily powerful patriarchal top-down autocracies of Adolph Hitler, Mohandas Gandhi, and Saddam Hussein. Their followers obeyed orders as if their life depended upon it. Moralistic criticism on past cases notwithstanding, under such circumstances it would be unthinkable on many levels to disobey the leader chief, and indeed such almost never happened.

In contrast, the orgeic matripolar path to and style of male leadership is opposite to the haremic patripolar pattern. Thus, orgeic culture is a matriarchal and nonviolent democratic commune, where age and wisdom are revered. Originally, all males in the camp nonviolently competed against each other in their courtship of the reproductively dominant female. She selected and retained each of them, but only after receiving their individual submission and pledge undying love to her and her children in the *motherland*. Because each male potentially had sex with all of females bearing offspring into the clan, each child was viewed as his own. Thus, he was also blood-bonded to the troupe by family loyalty.

While the queen attended to global goals, such as the long-range planning of the camp, a trusted prime minister consort, the temporarily dominant alpha-male, attended the important daily details. He arrived at his tentative conclusions by robust competition with the ideas of the other males in his parliamentary gang. This was an early form of bottom-up democracy. However here, all votes were not equal, but instead each weighted by the member's personal status within the troupe.

Attempts to govern mixed polaric groups, either under *only* autocracy or *only* democracy have resulted in centuries of conflict and hatred, and still are unfortunately underway. However, if patripolar autocratic order and constraint was the original thesis, and matripolar democratic freedom and chaos its antithesis, then the republican-democracies represent the synthesis of the two. Of the more recent solutions to governance of populations of mixed polarity, the presidential two party government of the US, a republican democracy informed by Greek democratic and French Revolution models, has been the most successful in creating a balanced golden mean, at least for brief periods

In the dialectic left "thesis", the citizens and society are represented in the US by the predominantly LM Democratic Party, while at the dialectic right "antithesis", wealthy land owners, businesses, and other vested power interests, are represented by the predominantly RM Republican Party. As both of these legitimate political party orientations fight vigorously for the advantage of their own interests, the pendulum swings back and forth, and the country as a whole staggers down the middle of the road. However, in the end only the authority of the president and his vice-president can lead them to a useful compromise. This compromise represents the dialectic "synthesis", the golden mean, whereby the nation is guided down the middle road of optimal survival. In a recent period of unparalleled prosperity, charismatic president Bill Clinton was a RM and vice-president Al Gore was a LM, thus inherently non-charismatic.

Now, with knowledge of the existence and effects of familial polarity sub-populations, it becomes theoretically possible logically to design new types of governments that can replace current global conflict with greater cooperation and mutual benefit within and across the polaric cultures.

In the American form of democracy, all votes are equal, including those of the females. This golden mean deviates from both the Orgeic (matripolar) hierarchal style of status where no male is equal, and the Haremic (patripolar) style of dominance where women's votes are fused with their husbands. Relevant to the latter is the case of LM former president Jimmy Carter, who with his RF wife, Rosalyn, threatened to leave the patriarchal Baptist Church because of their own belief that women are not naturally subservient to men. This is absolutely the case for matripolars. However, since most Protestants are patripolar (and Republican), many RM Baptist men and, interestingly, their LF women feel the reverse is true.

Occasionally it is advantageous to become that rare crossover from the opposite side. Thus, RM child-Jesus talking with the LM temple scholars was unusual, as was Islamic, sir-named RM Gandhi as a Hindu, the charismatic RM Kennedy's as Catholic, non-charismatic LM Jimmy Carter as a Baptist, and charismatic RM Bill Clinton as a Democrat, stealing fire from the RM Republicans.

Furthermore, to have a charismatic RM leader includes implicit acceptance of the Haremic inherently polygamous biology of a harem leader. Thus, there often will be adoring women with whom he must contend. In this regard, beloved RM president Kennedy (in spite of his lovely LF wife) said, he just could not help his ongoing gross promiscuity. Interestingly, the US presidents who have been rated as the best, historically usually have been charismatic RMs (84).

Observation #30: Familial Polarity Interspecies Conflicts Have Played a Major Role in History

Unrecognized Global Conflict between Patripolar and Matripolar Human Groups

From within mankind's present ignorance of Polarity, well-meaning LPs and RPs, tend to misunderstand each other. Because of opposite value systems, they commonly disagree, sometimes violently. For, example, the 1978 Bonn Summit Meetings were held between LP Jimmy Carter of the US and RP Helmut Schmidt, of Germany, to settle issues of mutual interest. Returning from each meeting, reporters on both sides released the debriefing information to the press in their respective countries. Both the Carter and Schmidt teams claimed quite different outcomes resulting from the same meeting. So different, that some of these men became inflamed, calling each other liars. By the time the meetings concluded, these individuals, if not nations, had grown to hold each other in deepest distrust. Yet, they both were teams of sincere, well-meaning allies, each aligned toward many of the same goals.

This illustrates the destabilizing effects that unrecognized polarity differences can produce even during peacetime international government interactions. If by accident, RM Bill Clinton had represented the US at that time, the outcome no doubt would have been much more positive, as it was for RM Kennedy earlier. This would not be because Clinton was any better than Carter was, but because the polarities of the two national leaders would have matched and not been crossed. They would both have been speaking the same language: RP-ese, which is considerably different from LP-ese! These two language styles were recognized earlier, but cast in a male-female framework (123-125).

In earlier less stable situations, such matter-anti matter contacts have led to repeated annihilations. In fact, most of written world history centers on these polaric conflagrations. Such remained as relatively local events of human misery, until the advent of the industrial revolution and the beginnings of modern war technology. From then, wars became increasingly massive slaughters on increasingly global levels, leading to the advent of the World Wars. In the First World War, patripolar Germany, Austro-Hungary and Turkey took on matripolar Britain, France, and Russia in clashes of higher mortality than the earth had yet seen.

After these two Giants picked themselves up and licked their millions of wounds, Round 2, the Second World War, followed between with self-same Haremic Axis giants against their same European Orgeic mortal enemy Allies. However, this time the chaos expanded. While the European Titans were preoccupied in mutual annihilation, Patripolar Japan went on a rampage of conquest and subjugation in Matripolar areas in Southeast Asia. Fortunately, for the outcomes of both of these world war rounds, a successfully governed, mixed-polarity country (US) was able to neutralize the champions, but at great cost.

Clearly, bringing permanent solutions to these escalating dialectic battles between universal Ying and Yang is of the highest priority. With the explosion of technology and the addition of Islamic and Northern Oriental patripolars to the conflagration, a Round 3 could eliminate all humans from the planet!

That the two human polarities have continued to resist interbreeding even in the present is indicated by the specific locations of repeated global unrest, violence, and genocide. These are usually found between the Eastern Hemisphere familial polarity striations (earlier in chapter) at the immiscible interfaces between two biologically different populations of opposite polarities (84).

Because of the existence of these striations, familial polarity also plays an unrecognized roll in global conflict. In many cases, these have been sites of violent conflict for centuries, sometimes for millennia. **Table 20** identifies many of these "Hot Spots of Violence" as recurring global sites of killing or genocide. Amazingly, 20 of 21 (95%) of these sites of repeated unrest were at the interface between populations of opposite familial polarity! That is, matripolar and patripolar soldiers were found to be on opposite sides of the battle lines. For example,

Table 20, Modern History: Matter vs. Antimatter at Interfaces between Polaric Populations

	French **vs.** Germans	
	Russians, Slavs **vs.** Germans	
	Jews **vs.** Germans	
MATRIPOLARS	English **vs.** Scottish	**PATRIPOLARS**
	Southern **vs.** Northern Irish	
FIGHTING FOR	Italians **vs.** Sicilians	FIGHTING FOR
	Spaniards **vs.** Moors	
MOTHERLAND	Spaniards **vs.** Basques	FATHERLAND
	Jews **vs.** Arabs	
	Serbs **vs.** Albanians	
"MOTHER!!!"	Russians **vs.** Chechnians	"FATHER!!!"
	Armenians **vs.** Turks	
"MOTHER !!!"	Indians **vs.** Pakistanis	"FATHER !!!"
	Indians **vs.** Sieks	
"I'M DYING !!! "	Hutu Farmers **vs.** Watutsi Warriors"	
	South **vs.** North Korea	"SAVE ME!!!"
"SAVE ME!!!"	South **vs.** North Vietnam "	
	Philippinos **vs.** Moros	I'M DYING!!!
	Romans **vs.** Greeks	
	*Sri Lankan Sinhalese **vs.** Indian Tamils	

* The one case of these 21 examples that was not Cross Polar

the Shia vs. Sunni Islamists; the Jews vs. Palestinians; the Serbs vs. Albanians; the Russians vs. Chechnyans; the Slavs and French vs. Germans; the English vs. Scottish; the Southern Irish vs. the Northern Irish; the Spaniards vs. Basques and Moors; the Italians vs. Sicilians; the

Indians vs. Pakistanis and Sieks; and the Hutu farmers vs. Watutsi Warriors.

This global situation has arisen in our ignorance of Familial Polarity. Such startling observations add a new dimension to achieving our goal of global peace. Certainly these facts can be utilized to enable us to evolve to the next level of human evolution: from the killing of **Figure 22** to nonkilling!

Figure 22, A History of Endless, Needless, Global Conflict and Slaughter

Matripolar Slavic Cro-Magnons   Patripolar Albanian Neanderthals*

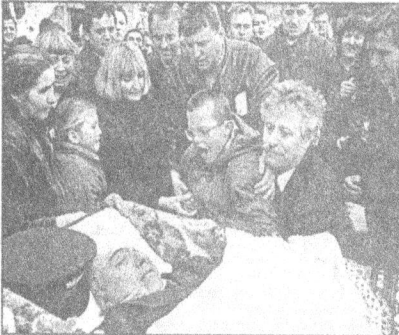

Associated Press

The young son of a Serb policeman, killed in a gun battle with ethnic Albanians, screamed in anguish at his father's funeral yesterday.

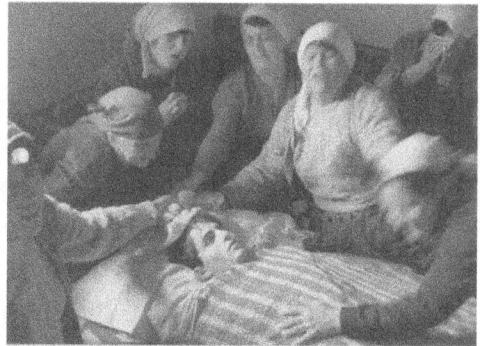

DRAGODIL, KOSOVO, OCTOBER 28, 1998: A group of ethnic Albanian women weep over the body of Ali Murat Pacarist, a 36 year old Kosovo Liberation Army soldier killed while trying to defuse a Serbian booby trap.

*Note the prominent brow ridge on the Albanian youth on the right. Ancient Neanderthal caves are nearby.

Hemisity Sampling of Current Distributions of Matripolar and Patripolar Populations:

In studies of familial polarity, large amounts of data have accumulated from the measurement of hemisity of over 1000 individuals within the community of the

University of Hawaii (UH) at its research campus in Manoa where over 20,000 multiethnic students are enrolled (84). These preliminary data, tabulated in **Table 21**, indicated that individuals, drawn from specific ethnic or geographic locations of diverse populations of familial polarity, varied greatly in their relative matripolar to patripolar ratios (M/P Ratio). For example, individuals of Germanic, Middle East, or the Northern Orient origins appeared to be predominantly patripolar in composition with low M/P Ratios. In contrast, many Southern European and some, but not all Southeast Asian populations appeared to be predominantly matripolar in hemisity, each having high M/P Ratios.

To understand further the nature of the genetic complexity present within races, these data, together with extensive ethnographic analysis the familial polarity of the cultures and religions of certain of these sub-populations, have been combined into a preliminary and tentative familial polarity assessment within some of the Old World countries. **Table 22** summarizes this analysis.

Immediately visible is the presence of immiscible competing matri- and patripolar population elements with their matri- and patripolar cultures and religions whose distribution appears more ancient than current national or even racial boundaries. Clearly, ignorance of familial polarity kills.

In contrast, **Table 23** illustrates how recognizing the existence of familial polarity and understanding its significance can reduce conflict within each of ten universe levels. Clearly, until we recognize the multilevel existence of familial polarity, it will continue to take its

Table 21: Ethnicity and Polarity of Subjects from the University of Hawaii Community

Country of Origin	Patripolars	Matripolars	M/P Ratio
Scotland	31	1	0.03
Northern Ireland	12	0*	0.08
England	8	44	6
Southern Ireland	3	24	8
French Canada	1	14	14
Western Canada	19	4	0.21
Germany	47	12	0.25
France	3	29	10
Spain	2	32	16
Italy	2	28	14
Sicily	16	0	0.06
Hungary	9	1	0.11
Poland	0	18	18
Russia	2	16	8
American Indians	15	2	0.13
Mexico	3	33	11
Hawaii	51	23	0.45
Samoa	27	5	0.18
Tonga	18	2	0.11
N. Philippines	3	41	14
S. Philippines	18	1	0.06
Okinawa	29	0	0.03
Japan	65	32	0.49
South China Cantonese	3	40	13
North China, Mandrin	50	4	0.11
North Korea	10	2	0.20
South Korea	1	28	14
Thailand	0	11	11
Egypt	11	0	0.09
Israel	2	41	10
Palestine	14	1	0.09
Pakistan	15	1	0.07
Indian Hindu	0	28	14
Indian Seik	12	0	0.08
Bangladeshi	8	2	0.23
Black American	3	49	16
Subjects: n=1089	514	575Ms	
Average M/P Ratio			1.12

* = Zero values were arbitrarily assigned the value of one. *Italics = Patripolar*

Table 22. Familial Polarity Estimates: Eastern Hemiphere

Country-Location	PATRI-RM Ethnicity	POLAR RM Religion	MATRI-LM Ethnicity	POLAR LM Religion
France	Huguenot	Protestant	French	Roman Catholic
Ger-Aust-Swi	Teuton	Protestant	Slav, Mediterr	Roman Catholic
Britain	Scottish	Protestant	English	Episcopal (Cath)
British Isles	Northern Irish	Protestant	Southern Irish	Roman Catholic
Italy	Sicilian-Greek	Greek, Pagan	Italian-Roman	Roman Catholic
Spain	Catalonian	Protestant	Castilian	Roman Catholic
Morocco	Moor	(orig.)	Berber	Animistic
Greece	Greeks	Islamic	Slavs	Greek Orthodox
Russia	Chechnya etc.	Greek Pagan	Russian	East. Orthodox
Yugoslavia	Albanian, etc.	Islamic	Serbia	East. Orthodox
Turkey	Turkish	Islamic	Armenian	East. Orthodox
Israel	Palestinian	Islamic	Jewish	Judaism
Arabia, Iraq, Iran, Syria, Iran Lebanon, Egypt, Afghanistan,	Persian Arab, etc.	Islamic	Opposition is	not well tolerated!
Africa	Nilotic hunter-herdsmen, Watutsis, etc.	Islamic	Bantus Farmers, Hutus, etc.	Catholic, pagan
Indian sub-continent	Pakistani, Bangladeshi	Islamic	Indian	Hindu
Southeast	Myanmar	Islamic	Thai,Sri Lakn	Buddhist
Asia	S.Filipino	Islamic	N. Filipino	Roman Catholic
	Moro	Islamic	Tagalog	
China	Mandarins	Confucianism	Cantonese	Taoism
Korea	North Korea	Confucianism	South Korean	Buddhist
Viet Nam	N Vietnamese	Buddhist	S.Vietnamese	Roman Catholic
Australasia	Aborigines, Polynesians, Papuans	Pagan	Melanesian, Micronesian	Animistic

Table 23: Understanding Familial Polarity can Reduce Conflict at Ten Universe Levels

This knowledge:

1. Resolves polarity-based motivation and identity conflicts within **myself**.
    Seeing my polaric identity as perfect, I know that I belong to my global family.
2. Resolves polarity-based conflicts in **my spiritual life**.
    Knowing my hemisity clarifies that my kind of religious experience is normal.
3. Avoids polarity-based conflicts within **my nuclear family**.
    Understanding the polarity needs of each, I can create synergy.
4. Deals with polarity-based conflicts within **my extended family**.
    Recognizing the existence of LP sensitivity and RP intensity removes misunderstandings.
5. Reduces polarity-based conflicts with **my neighbors** (Western Hemisphere) Identifying polarity differences of neighbors helps me to accept them as family again.
6. Avoids polarity-based conflicts within **my community** (Eastern hemisphere). Patripolar and matripolar family groups are inherently different with different needs.
7. Lowers polarity-based conflicts within **my town or city**.
    Unique patripolar and matripolar group strong and weak points become complementary.
8. Reduces polarity based conflicts within **my state**.
    Dyadic pendulum extremes of opinion become easier to recognize and stabilize.
9. Prevents polarity based conflict within **my nation**.
    Knowledge of reality, the origin, and nature of life brings wisdom to social policy.
10. Avoids polarity based international conflict within **my world**.
    Recognition that moving from a selfish national level to family of nations level interaction transforms competitive politics into a cooperative compeimentary global network of peace and prosperity.

toll. This begins through ignorance inflicting emotional trauma upon our young children, especially in hybrid families where child rearing often is crosswired. Such leads to permanent developmental arresting and self-loathing. This causes stress-sensitization, neurosis, and reward-hunger (drug seeking) in adulthood. Intense drug demand to self-medicate psychic pain produces personal and global corruption, waste, violence, killing, and war. These themes are further developed in *Neuroreality: A Scientific Religion to Restore Meaning* (93).

CHAPTER 11

THE TWO HUMAN SPECIES: FAMILIAL POLARITY

CHAPTER 11. SUMMATION: A CONSCIOUSNESS AWAKENING

A SUMMARY CHAPTER

In this day of increasing global communication, we are now confronted with situations that could conveniently be ignored and avoided in the past. Homosexual marriages involve voters of entire states. Lesbian and gay partners are rearing thousands of children. Millions are attracted to the dark side, body piercing, body sculpting, and tattooing. Exhibitionistic gothic type rock stars attract millions of avid followers. Homosexuals are becoming politically prominent. They speak directly to us daily as TV news anchors and other persons. There is massive public targeting of our ubiquitous pedophiles. The bizarre behavior of schizophrenics makes headlines.

In addition, these strange but often highly talented people must undergo a down side that is becoming increasingly obvious. Their inability or unwillingness to meet societal norms results in their being bullied and ostracized. This results in their suffering, despair, abuse of intoxicating drugs, desire for revenge against society, or solution by suicide.

Who are these strange people? Where are they coming from? Where they born this way? If so why? What should be done? This brief summary gives the short version of the answers developed in this book.

The Introduction documents the extent of the present global situation, giving facts and figures about the extent of many of these different nonstandard cross-wired human groups. This book covers dyslexics, trans-heterosexuals, trans-homosexuals, cis-homosexuals, gothic dark side types, and pedophiles and schizophrenics.

Chapter 1 introduces the life-changing topic of "Hemisity." It also enables you to learn your own hemisity and that of people important to you. That is, whether yourself, your friends, and relatives belong to the very different right brain-oriented (R) or left brain-oriented (L) hemisity subtypes. Four published hemisity questionnaires are provided in the Appendix for this purpose.

The chapter begins by reviewing what has recently been discovered about the existence of inherent right brain and left brain behavioral differences. In the 1980s, psychologists and psychiatrists rejected research on behavioral laterality, called hemisphericity, as premature. They effectively made it a politically incorrect subject. They felt forced to do so because they lacked proper definitions and the quantitative methods to distinguish a person's true hemisphericity at that time, to assess the ethnic laterality and other claims of that time.

Since then, the author has discovered, developed, and published a number of quantitative biophysical hemisity methods. These supported the idea that we are all either right or left brain-oriented persons (RPs or LPs), with no "in betweens". Further, it was found that RPs and LPs have inherently different perspectives on life. Thirty specific items have been found thus far where RPs and LPs differ in their preferences. Of importance to us here is that RPs (M or F) are big picture, lumpers, and dominant marital partners. In contrast, the LPs (M or F) are important details, splitters, who are supportive and complimentary to their RP partners,

You have now learned how to spot hemisity subtypes in the people around you and in the media. The estimated hemisity of some historical and current public figures is given. By the end of this chapter, you began to see the obviousness of hemisity and of how it has always surrounded us.

Chapter 2 relates research regarding the hemisity of parental partners. There are four possible pairings: the two complimentary pairs: RM-LF and RF-LM, and the two like-like pairs: RM-RF and LM-LF (in terms of male or female, right or left brain-orientation). Recent research (84) demonstrates that in most samples, 67% human reproductive pairs consist of partners of opposite, complementary hemisity. That is, in courtship, marriage, and reproduction, "opposites attract" is the rule in terms of hemisity. Since one's right or left hemisity subtype is inborn, and unchangeable, this means that there must be at least two pre-racial human species: the patripolar RM-LF lineages and the matripolar RF-LM lineages. This has great historical significance and is be shown to be at the core of ongoing global conflicts. At this point, you should be able to determine your family polarity.

In Chapter 3 we reached the heart of our subject: the unrecognized production of cross-wired hybrid children born from the mating between the matripolar and patripolar human species. For those 33%, like-like marital pairs, RM-RF and LM-LF, their children will be hybrids because both parental pairs are of mixed familial polarity. That is, a RM-RF pair consists in a patripolar RM married to a matripolar RF. Similarly, the LM-LF pair is a cross between a matripolar LM and a patripolar LF. Cross-wired hybrid offspring are inevitable. Like many other known cross species matings, these hybrid offspring can have several possible reproductive or other differences or anomalies.

Chapter 4 was devoted to the origin and properties of developmental dyslexics, whose abundance in the population has been estimated to be as high as one in five. Normal wild types are born with their memory system on the same side as their Executive. This gives them, not only photographic memory, but also great the rapidity of recall and processing. This makes them excellent speakers and leaders.

In contrast, developmental dyslexics, all of whom are RPs, have their mind cortical memory system born unnaturally on the left side of their brains away from and inaccessible to their right sided Executives. They must retrieve their memory from the left side, resulting in reversed mirror images being imported to their Executive on the right, which must be corrected. Further, not everything available in their memory system comes across from their left memory bank, for example, names and numbers do not. This makes dyslexics slower thinkers who must make additional effort to memorize their own phone numbers. They do have limited sub-cortical short and long-term memory elements. Furthermore, when they become frightened, transfer of information from their memory base in the LH becomes inhibited, as in a mental block. This causes their mind to become blank, making them temporarily stupid with stage fright. Once the fear has passed, they regain their database, composure, and intelligence, which often is high. Developmental dyslexics are the RPs offspring from RM-RF marriages. No LP offspring have been found with developmental dyslexia. Millions of developmental dyslexics are now suffering because they were born that way. Early detection and training of dyslexics can enable them to excel at life. Some have become geniuses.

In Chapter 5 we found that LP offspring from the above RM-RF marriages have cross-wired reversed sexual identities, while remaining heterosexual (or bisexual). They have superb memories. There are an enormous number of these unrecognized adult trans-heterosexual offspring among us. They surround us as the softer more intellectual, artistic uncles, or the aunties with short hair and coarser textured clothes. Since awareness of their existence has not become part of our culture, they remain unseen and unnamed. Perhaps the closest terms for male and female Trans-Hets, are "wimps" or "jocks", kindly said out of hearing range. As in, "He doesn't drive trucks. She does." They are the hybrid capable of bisexuality.

In Chapter 6 we came to the hybrid offspring of LM-LF cross-polar couples, one fourth of which have been found to be homosexual. Cis (same side) and trans (opposite side) are terms derived from chemistry nomenclature. In this application, a cis-homosexual accepts his body sex, but prefers partners also of his same body sex. A trans-homosexual rejects his body sex, but prefers partners of his same rejected body sex. Obviously, these people have sustained irreversible developmental failures, resulting in unnatural or altered cross-wiring regarding partner preferance, and in the trans-homo case, of their own sexual identity. This results in non-standard behavior on their part, resulting in rejection, bullying, and ostracism from the wild types. Yet, they were born this way. It is natural to them and cannot be changed. They have begun to organize and protect themselves, especially against bullying as children and teenagers

Chapter 7 addressed the charged subject of pedophilia. Pedophiles are familial polarity hybrids who are born cross-wired to be unnaturally attracted to helpless children as sex objects, instead of to adults who can defend themselves. As Freud and others have noted, the primary sex drive to reproduce is inbuilt in all life forms. It incessantly and powerfully overrides everything. During maturity, it appears that scarcely an hour goes by without conscious and subconscious calculations being made sexual accessibility of target reproductive objects, originally of the opposite sex.

When those sex objects are someone else's children, a murderous hatred toward the molester automatically arises in those children's parents. Further, the children molested never recover their normal sexuality. Yet, for the pedophiles, their nagging hour-by-hour cross-wired instinct to reproduce with children cannot be abated. As Familial Polarity hybrids, they were born this way. This specific family polarity hybrid needs to be identified and treated with insight and caution.

Chapter 8 provided a new clarity on the origin and nature of schizophrenia. This disease attacks perfectly normal young adults, and literally digests and destroys their brains so they cannot tell reality from fantasy and cannot remember long enough to plan or execute anything. These undeserving victims become the epitome of persons with a broken brain who are completely helpless. This is yet another example of hybrid cross-wiring in utero, only one that expresses itself 20 years later in the final stages of brain maturation. Again, what crossing between which familial polarity hybrids needs to be determined, so that such disastrous matings can be avoided.

In Chapter 9, the strange subject of the LxP and Gothic and Dark Sides came to the fore. It's as if these people know that they are born not "standard issue", and that their drives and attitudes are clearly unnatural, but also absolutely unchangeable. This brings a self-revulsion and detestation of their nature that results in their developing contempt for religion and life replaced by an attraction to horror and death. This is not only manifest by mutilation of their bodies by cuts, metallic implants, and deforming anatomical inserts, but also by dressing as dead spirits in black, blood red, and white, the production of angry, discordant music, and the desire to seek revenge upon lucky normals, especially those whom ignorantly bully them. They grasp the hopelessness of their errors in cross-wiring and see suicide as a viable alternative, especially if they can kill their tormentors in the process, as in the Columbine Killings. Hopefully, as the Goth subculture develops, the motivation driving such antisocial events can be discharged in less harmful directions. In contrast, suicide bombers, jihadists and fanatics tend to be hyper-religious.

Chapter 10 went beyond cross-breeding to review the past, present, and future consequences to society of the existence of the two human species of familial polarity from a new perspective . Presently there is multi-leveled conflict and suffering occurring due to ignorance of familial polarity. This is unnecessary.

Chapter 11 summarized this book and now lists the evidence supporting the existence of two human species. The author believes that other interpretations of these observations will not fit the data as parsimoniously as the two human species hypothesis.

Data-Based Observations

1: The Hemisity of Spouses: Opposite Hemisities Commonly Attract

2: Right-Brained Spousal Dominance Within the Four Possible Hemisity Couple Types

3: The Hemisity of Children from the Four Hemisity Couples: Fixed vs. Random

4: Hemisity: Corpus Callosum Size Differences and Cingulate Cortex Size Laterality

5: Evidence that Hemisity is Inherited

6: Opposite Reproductive Strategies: Foundation of Endless Conflict

7: Survival of the Fittest Competition Occurs at Different Universe Levels in the Two Species

8: Opposite Courtship and Child Rearing Behaviors Exist

9: Existence of Contrasting Behavioral, Cultural, and Institutional Differences between the Human Polarities

10: Anatomical Differences Between the Two Species Exist

11: Terrestrial Life Has Originated More Than Once

12: There Appear To Be At Least Two Trees of Life

13: Two Opposite Reproductive Strategies Exist in the Apes, Our Closest Living Primate Relatives

14: Throughout the Archeological Record Matri and Patripolar Species of Hominids Existed in Parallel

15: Patripolar and Matripolar Lineages Have Made It Into The Present

16: Crossbreeding Produces Hybrids With Unnatural Neural Cross-wirings, Causing Several Behavioral Abnormalities, and Maladies.

17: It's No Longer Boy meets Girl, But 16 Kinds of Boy Meets 16 Kinds of Girls!

18: All Right-Brained Children From R-R Cross-breeding Parents Were Dyslexic

19: All Dyslexics Were Right Brain-Oriented RxPs

20: In Dyslexics, the Cortical Memory Module is Displaced to the Left Hemisphere

21: Fear Causes Dyslexics to Choke with Mental Blocks

22: LxP Hybrids Are Trans-Heterosexuals With Reversed Sexual Identities

23: In Trans-Heterosexuals, the Emotion Side of the Brain is Reversed to the Left Side

24: In Trans-Heterosexuals, Mirror Tracing Results are Opposite to Their Other Hemisity Test Results.

25: LxPs Have the Potential To Be Bisexual

26: Cis-Homosexuals are Familial Polarity Hybrids

27: Trans-Homosexuals Are Familial Polarity Hybrids

28: Homosexuals Can Also Be Dyslexic

29: Cause of Pedophilia: Sexual Drive Becomes Miswired to an Immature Reproductive Target

30: Hybrid Familial Polarity Schizophrenia Model

31: Most Goths are Familial Polarity Hybrids.

32: Familial Polarity Interspecies Conflicts Have Played a Major Role in History.

It is hoped that the introduction of the new context of familial polarity with its two human species will result in a clarification of many previously opaque problems of human existence. It is believed that the application of this information can reduce human suffering and increase global harmony and personal satisfaction.

CHAPTER 11

TWO HUMAN SPECIES: FAMILIAL POLARITY

REFERENCES

[1] Abiotiz F, Scheibel AB, Fisher RS, & Zaidel E. (1992). Fiber composition of the human corpus callosum. *Brain Research*, **598**, 143-152.

[2] Amen DG. (1998), *Change your Brain. Change your Life,* Three Rivers Press, NY.

[2b] Auel JM (1980) The Clan of the Cave Bear, Bantam, N.Y.

[3] Balter M & Gibbons A. (2000). A glimpse of human's first journey out of Africa. *Science,* **288**, 948-950.

[3b] Barinaga M. (1994). Cell Suicide: By ICE, not Fire. Science, **236**, 754-756.

[4] Beaumont G, Young A, & McManus IC. (1984). Hemisphericity: A critical review. *Cognitive Neuropsychology,* **1**, 191-212.

[5] Benefit BR & McCrossin ML. (1997). Earliest known Old World monkey skull. *Nature,* **388**, 368-371.

[6] Bishop KM & Wahlsten D. (1997). Sex differences in the Human corpus callosum: Myth or reality? *Neuroscience and Biobehavioral Reviews,* **21**, 581-601.

[7] Bogen JE, & Vogel PJ. (1962) Cerebral commissurotomy: A case report. *Bulletin of the Los Angeles Neurological Society,* **27**, 169-174.

[8] Bogen JE, & Fisher ED, Vogel PJ. (1965). Cerebral commissu rotomy: A second case report. *Journal of the American Medical Association,* **194**, 1328-1329.

[9] Bogen J E. (1969). The other side of the brain. I. Dysgraphia And dyscopia following cerebral commissurotomy. *Bulletin of the Los Angeles Neurological Society,* **34**, 75-105.

[10] Bogen, JE. (1969). The other side of the brain II. An appositional mind. *Bulletin of the Los Angeles Neurological Society,* **34**, 135-162.

[11] Bogen JE, & Bogen GM. (1969). The other side of the brain. III. The corpus callosum and creativity. *Bulletin of the Los Angeles Neurological Society*, **34,** 191-219.

[12] Bogen JE, DeZure R, TenHouten WD, & Marsh, J. F. (1972). The other side of the brain. IV. The A/P ratio. *Bulletin of the Los Angeles Neurological Society, 37,* 221-235.

[12b] Bottini GR, Corcoran R, Sterzi E, Paulesu P,. Schenone P, Scarpa RS, Frackowiak HY & Frith C.D. (1994). The Role of the Right Hemisphere in the Interpretation of Figurative Aspects of Language: A Positron Emission Tomography Activation Study. *Brain,* **117,** 1241-1253.

[12c] Brown D, 2003. *The Da Vinci Code,* Doubleday, N.Y.

[13] Brown W S, Larson E B & Jeeves M A. (1994). Directional asymmetries in interhemispherici transmission time: evidence from visual evoked potentials. *Neuropsychologia* 32, 439-448.

[14] Brown HD & Kosslyn SM. (1993). Cerebral lateralizeation. *Current Opinion in Neurobiology,* **3**, 183-186.

[15] Callum, MS. (2007). Postadolescent brain development: A disconnect between neuroscience, emerging adults, and the corrections system. *Wisconsin Law Review*, 74010/26/2007 12:26 PM, 730-758.

[16] Carter CS, Botvinick MM & Cohen JD. (1999). The contribution of the anterior cingulate to executive processes in cognition. *Review of Neuroscience,* **10**, 49-57.

[17] Castillo RJ. (1997). *Culture and Mental Illness.* Brooks-Cole, Pacific Grove, CA.

[18] Clark S, Karftsik R, Van der Loos H, Innocenti GM. (1989) Forms and measure of adult and developing human corpus callosum: Is there sexual dimorphism? Journal of Comparative Neurolology, **280**, 213-230.

[19] Clarke JM & Zaidel E. (1994). Anatomical-behavioral relationships: corpus callosum morphometry and hemispheric selection. *Behavior and Brain Research,* **64**, 185-202.

[20] Corbera X, Grau C. (1993) Diurnal type and hemisphericity asymmetry. Cortex, **29,** 519-528.

[21] Crowell DH, Jones RH, Kapuniai LE & Nakagawa JK. (1973). Unilateral cortical activity in newborn humans: An early index of cerebral dominance. *Science,* **180**, 205-208.

[22] Culotta, G. (1995). Asian hominids grow older. *Science,* **270**, 1116-1117

[23] Daglish MR, Weinstein A, Malizia AL, Wilson S, Melichar J K, Britten S, Brewer C, Lingford-Huges A, Myles JS, Grasby P, & Nutt DJ. (2001). Changes in regional cerebral blood flow elicited by craving memories in abstinent opiate-dependent subjects. *American Journal of Psychiatry,* **158**, 1680-1686.

[23b] Darwin C. (1871). *The Descent of Man and Selection in Relation to Sex.* Prometheus Books. N.Y.

[24] Davidson RJ & Hugdahl K. (1995). *Brain Asymmetry,* MIT Press.

[25] Dawkins R. (1990). *The Selfish Gene,* Oxford University Press.

[25b] DeBruine L, Jones BC, Crawford JR, Welling L & Little AC. (2010). 'The health of a nation predicts their mate preferences: Cross-cultural variation in women's preferences for masculinized male faces'. *Proceedings of the Royal Society of London. B, Biological Sciences, vol 277,* 2405-2410.

[26] DeMeo J. (1998). *Saharaisa: The 4000BCE Origins of Child Abuse, Sex-Repression, Warfare, and Social Violence in the Deserts of the Old World.* Orgone Biophysical Research Lab, Ashland, OR.

[27] Descartes R.(1637). *La dioptrique:* In: Discours de la me thode. Leiden, Ian Maire, In: Adam & Tannery, Vol VI, Paris ,CNRS/vtn, 1964-74.

[28] Devinsky O, Morrell MJ, & Vogt BA.(1995). Contributions of the anterior cingulate cortex to behavior. *Brain,* **18**, 297-306.

[29] DeWaal F. & Lanting F. (1997). *Bonobo: The Forgotten Ape.* University of California Press, San Francisco.

[30] Diamond JM. (1991). The earliest horsemen. *Science,* **350**, 275-276.

[31] Edwards B. (1993). *Drawing on the Right Side of the Brain.* Harper-Collins. NY.

[32] Eide BL & Eide FF. (2011), *The Dyslexic Advantage.* Hudson Street Press, NY.

[33] Eisler R. (1987). *The Chalise and the Blade.* Harper-Collins, NY.

[34] Eisler R. (1995). *Sacred Pleasure.* Harper-Collins NY.

[35] Fink GR, Halligan PW, Marshall JC, Frith CD, Frackowiak RSJ & Dolan RJ. (1996). Where in the brain does visual attention select the forest and the trees? *Nature,* **382**, 626-628.

[36] Finkelhor D. (1984). Sexual Abuse as a Moral Problem. in *Child Sexual Abuse: New Theory and Research* (London: The Free Press, p 14.

[37] Finkelhor D. (1990). Early and Long-Term Effects of Child Sexual Abuse: An Update, *Professional Psychology: Research and Practice,* **21**, 328.

[38] Flaherty R. (1922) Nanook of the North. Silent film by Robert J. Flaherty.

[39] Fornito A, Yucel M, Wood S, Stuart GW, Buchanan J, Proffitt T, Anderson V, Velakoulis D, & Pantelis P. (2004). Individual differences in anterior cingulate /paracingulate morphology are related to executive functions in healthy males. *Cerebral Cortex,* **14**, 424-431.

[40] Fossy D. (1983). *Gorillas in the Mist.* Houghton Williams, NY..

[41] Furuichi T. (1992). Prolonged estrus of females and factors influencing mating in a wild group of bonobos (Pan paniscus) in Wamba, Zaire. In Topics in Primatology, vol 2; *Behavioral Ecology and Conservation,* ed. N. Itaigowa, Y. Sugiyama, GP. Sackett, & RKR. Thompson, 179-190. Univeristy of Tokyo Press.

[42] Galaburda AM. (1991). Asymmetries of cerebral neuroanatomy. *Ciba Foundation Symposia* **162**, 219-226; discus

sion, 226-233.

[43] Gazzaniga MS. (1967). The Split Brain in Man. *Scientific American,* **217**, 24-29.

[44] Gazzaniga, M.. S., (1989). Organization of the human brain. *Science,* **245**, 947-952.

[45] Gazzaniga, M. S., (2000). Cerebral specialization and interhemispheric communication; Does the corpus callosum enable the human condition? *Brain,* **123**, 1293-1326.

[46] Gazzaniga M, Bogen JE & Sperry RW. (1962). Some functional effects of sectioning the cerebral commisures in man. *Proceedings of the National Academy of Sciences, USA,* **48**, 1765-1769.

[47] Gazzaniga MS, Bogen JE, Sperry RW. (1967). Dyspraxia following division of the cerebral commissures. *Archives of Neurology,* **16**, 602-612

[48] Giedd JN, et al. (1999). Brain development during childhood and adolescence: a longitudinal MRI Study. *Nature Neuroscience* **2**, 861-3.

[49] Galdikas BMF. (1995). *Reflections of Eden.* Little and Brown, NY..

[50] Gibbons A. (1997). Bone sizes trace the decline of Man (and Woman). *Science,* **276**, 896-897.

[51] Goodall J. (1990). *Through a Window,* Houghton, Mifflin, NY.

[52] Gray J. (1992). *Men are from Mars, Women are from Venus.* Harper Collins, N.Y.

[53] Gootjes L, Bouma A, van Strien JW, Van Schiljndel R, Barkhof F, & Scheltens P. (2006). Corpus callosum size correlates with asymmetric performance on a dichotic listening task in healthy aging but not in Alzheimer's disease. *Neuropsychologia,* **44**, 208-21.

[54] Hasegawa I, Fukushima T, Ihara T & Miyashita Y. (1998). Callosal window between prefrontal cortices: Cognitive Interaction to retrieve long-term memory. *Science,* **281**, 814-818.

[55] Henry JP & Wang S. (1998). Effects of early stress on affilia

tive behavior. *Psychoneuroendocrinology,* **23**, 863-895.

[56] Hines M, Chiu L, McAdams LA, Bentler PM & Lipcamon J. (1992). Cognition and the corpus callosum: verbal fluency, visuo spatial ability, and language lateralization related to mid sagittal surface areas of callosal subregions. *Behavioral Neuroscience,* **106**, 3-14.

[57] Holloway RI. (1985). The poor brain of Homosapiens neanderthalensis: See what you please. In: *Ancestors: The Hard Evidence*, ed. by E. Delson, pp. 319-324, Alan R. Liss, NY.

[58] Holloway RL, Anderson PJ, Defendini R, & Harper C. (1993). Sexual dimorphism of the human corpus callosum from three independent samples: relative size of the corpus callosum. *American Journal of Physical Anthropology*, **92**, 481-492.

[59] Hutsler JJ, Loftus WC, Gazzaniga MS.(1998). Individual variation of cortical surface area asymmetries. *Cerebral Cortex*, **8**, 11-17.

[60] Ide A, Dolezal C, Fernandez M, Labbe E, Mandujana R, Montes S, Segura P, Verschae G, Yarmuch P, & Abiotiz F. (1999). Hemispheric differences in variability of fissuaral patterns in parasylvian and cingulate regions of human brains. *Journal of Comparative Neurology*, **410**, 235-242.

[61] Ivry RB, & Robertson LC.(1998). *The two sides of perception.* MIT Press, Cambridge, MA .

[62] Jager G. & Postma A., (2003). On the hemispheric specialization for categorical and coordinate spatial relations: A review of the current evidence, *Neuropsychologia,* **41**, 504-515.

[63] Janke, L & Steinmetz, H., (1994). Interhemispheric transfer time and corpus callosum size. *Neuroreport,* **5**, 2385-8.

[64] Janke L, Staiger JF, Schlaug G, Huang Y, Steinmetz H. (1997). The relationship between corpus callosum size and forebrain volume. *Cerebral Cortex*, 7:48-56.

[65] Kanno T. (1992). *The Last Ape: Pygmy Chimpanzee Behavior and Ecology.* Stanford University Press, Palo Alto, CA.

[66] Kerr R. (2000). A victim of the Black Sea Flood found.

Science, **289**, 2021.

[67] Kertesz A, Polk M, Howell J, & Black SE. (1987).
Cerebral dominance, sex, and callosal size in MRI.
Neurology, **37**, 1385-8.

[68] Kimura, D. (1967). Functional asymmetry of the brain in
dichotic listening. *Cortex*, **3**, 163-178.

[69] Kosslyn SM. (1987). Seeing and imagining in the cerebral
hemispheres: A computational approach, *Psychological
Review,* **94**, 148–175.

[70] Kosslyn SM, Koenig O, Barrett A, Cave C, Tang J & Gabrieli
JDE. (1989). Evidence for two types of spatial representa-
tions. *Journal of Experimental Psychology: Perception and
Perfomance,* **15**, 723-35.

[71] Kosslyn SM, Chabris CF, Marsolek CJ & Koenig O. (1992).
Categorical versus coordinate spatial relations: computational analyses and computer simulations. *Journal of Experimental Psychology: Human Perception and Performance,* **18**, 562-577.

[72] Kunuk Z (2001)Atanarjuat: The Fast Runner, Canadian film,
produced by Zacharias Kunuk.

[73] Lamb MR, Robertson LC, & Knight RT. (1990). Component
mechanisms underlying the processing of hierarchically organized patterns: Inferences from patients with unilateral
cortical lesions. *Journal of Experimental Psychology:
Learning, Memory, and Cognition,* **16**, 471-483.

[74] Lang J & Ederer M. (1980). Shape and size of the corpus
callosum and septum pellucidem. *Gegenbaurs Morphologie
Jahrbuch,* **126**, 949-958.

[75] Lev-Yadun S, Gopher A & Abbo S. (2000). The Cradle of
Agriculture. *Science,* **288**, 1602-1603.

[76] Lockhorst GJ. (1985). An ancient Greek theory of hemispheric specialization. *Clio Medica*, **17**, 33-38.

[77] Libet B, Gleason CA, Wright EW, Pearl DK. (1983). Time of
conscious intention to act in relation to onset of cerebral activity (readiness potential): The unconscious initiation of a
freely voluntary act. *Brain*, **106**, 623-642.

[77b] Little JR, (1989) *Contemporary female bisexuality: A psychosocial phenomenon,* Unpublished doctoral dissertation.

[78] Ludders E, Rex DE, Narr KL, Woods RP, Jancke L, Thompson PM , Mazziotta, & JC, Toga, AW. (2003). Relationship between sulcal asymmetries and corpus callosum size: Gender and handedness effects. *Cerebral Cortex* , **13**, 1084-1093.

[79] MacLean PD. (1978). Effects of lesions of globus pallidus on species-typical display behavior of squirrel monkeys. *Brain Research,* **149**, 175-196.

[80] Margulis L & Sagan D. (1991). *Mystery dance: On the evolution of human sexuality.* Simon and Schuster, N.Y..

[80b] McConaghy N. (1998). Paedophilia: A review of the evidence. *Australian and New Zealand Journal of Psychiatry. 32*, 252-65, discussion 266-7.

[81] McGlone J.(1980) Sex differences in human brainasymmtry: A critical survey. Brain and Behavioral *Science*, **3**, 215-263

[82] Mitchell TN, Free LL, Merschhemke M, Lemieux L, Sisodiya SM, & Shorvon SD, (2003). Reliable callosal measurement: population normative data confirm sex-related differences. *American Journal of Neuroradiol ogy,* **24**, 408-410.

[83] Moore HDM, Martin M, & Birkhead T. (1999). No evidence for killer sperm or other selective interactions between human spermatozoa in ejaculates of different males in vitro. *Proceedings of the Royal Society*, Section B, 266, 2343.

[84] Morton BE. (2000). Unpublished data.

[85] Morton BE. (2001). Large individual differences in minor ear output during dichotic listening. *Brain and Cognition,* **45**, 229-237.

[86] Morton BE. (2002). Outcomes of hemisphericity questionnaires correlate with unilateral dichotic deafness. *Brain and Cognition,* **49**, 63-72.

[87] Morton BE. (2003a). Phased mirror tracing outcomes

correlate with several hemisphericity measures. *Brain and Cognition*, **51,** 294-304.

[88] Morton BE. (2003b). Two-hand line-bisection task outcomes correlate with several measures of hemisphericity. *Brain and Cognition* **51***,* 305-316.

[89] Morton BE. (2003c). Asymmetry questionnaire outcomes correlate with several hemisphericity measures. *Brain and Cognition*, **51,** 372-374.

[90] Morton BE. (2003d). Line bisection-based hemisphericity estimates of University students and professionals: Evidence of sorting during higher education and career selection. *Brain and Cognition* **52***,* 319-325.

[91] Morton BE & Rafto SE. (2006). Corpus callosum size is linked to dichotic deafness and hemisphericity, not sex or handedness. *Brain and Cognition*, **62**, 1-8.

[92] Morton BE & Rafto SE. (2010). Behavioral laterality advance: Neuroanatomical evidence for the existence of hemisity. *Personality & Individual Differences*, *49*, 34-42.

[93] Morton BE. (2011). *Neuroreality: A Scientific Religion to Restore Meaning, or How 7 Brain Elements Create 7 Minds and 7 Realities.* Megalith Books, Doral, FL.

[94a] Morton, B. E. (2012a). Right and left brain-oriented hemi sity subjects show opposite behavioral preferences. *Frontiers in Physiology.3:*407. doi:10.3389)./fphys.2012.00407.

[94b] Morton, B. E. (2013). Behavioral laterality of the brain: Support for the binary construct of hemisity. *Frontiers in Psychology,* 4 (doi: 10.3387/fpsyg.2013.00683).

[94c] Morton, B. E., Svard, L. & Jensen, J. (2014). Further Evi dence for Hemisity Sorting during Career Specialization. *Journal of Career Assessment,* 22 (2), 317-328.

[94d] Morton, B. E. (2017a). The Dual Quadbrain and Modular Consciousness. *Universal Journal of Psychology,* 5 (3), 157-166. doi: 10.13189/ujp.2017.050308

[95] Report of the National Reading Panel (2000). Teaching

Children to Read: An Evidence-Based Assessment of the Scientific Research Literature on Reading and Its Implications for Reading Instruction.

[95b] O'Kusky J, Strauss I, Kosaka B, Wada J, Li D, Druhan M, & Petrie J. (1988). The corpus callosum is larger with right-hemisphere cerebral speech dominance. *Annals of Neurology,* **24**, 379-383.

[96] Ornstein T.(1997). The right mind: Making sense of the hemis pheres. Harcort Brace, NY.

[97] Pakkenberg B, & Gundersen H J. (1997). Neoortical neuron number in humans: effect of sex and age. *Journal of Comparative Neurology* **384**. 312-320.

[98] Palline R, Agliotti S, Tassinari G, Berlucchi G, Colosimo C, & Rossi GF. (1995). Callosotomy for intractable epilepsy from bihemispheric cortical dysplasia. *The European Journal of Neurosurgery,* **132,** 79-86.

[99] Paus T. (2001). Primate anterior cingulate cortex: where motor control, drive, and cognition interface. *Nature Neuroscience, 2,* 417-424.

[100] Paus T, Tomaiuolo F, Otaky N, MacDonald D, Petrides M, Atlas J, Morris R, & Evans AC.(1996a). Human cingulate and paracingulate sulci: Pattern, variability, asymmetry, and probabilistic map. *Cerebral Cortex,* **6**, 207-214.

[101] Paus T, Otaky N, Caramanos Z, MacDonald D, Zijdenbos A, D'Avirro D, Gutmans D, Holmes C, Tomaiuolo F, Evans AC. (1996b). In vivo morphometry of the intrasulcal grey matter in the human cingulate, paracingulate, and superior-rostral sulci: hemispheric asymmetries, gender differences, and probability maps. *Journal of Comparative Neurology,* **376**, 664-673.

[102] Petrides M. (2000). Dissociable roles of mid-dorsolateral prefrontal cortex and anterior infero-temporal cortex in visual working memory. *Journal of Neuroscience, 20,* 7496-7503.

[103] Pfefferbaum A, Sullivan EV, Sean GC, Carmelli D.(2000). Brain structure in men remains highly heritable in the seventh

and eighth decades of life. *Neurobiology of Aging*, **21**, 63-74.

[104] Pollin W, Allen MG, Hoffer A, Stabenau JR & Hrubec Z. (2007). Psychopathology in 15,909 Pairs of Veteran Twins: Evidence for a Genetic Factor in the Pathogenesis of Schizophrenia and Its Relative Absence in Psychoneurosis *American Journal of Psychiatry* **126**, 597-610

[105] Robertson LC & Lamb MR. (1991). Neuropsychological contributions to theories of part/whole organization. *Cognitive Psychology*, **23**, 299-330.

[106] Sandweiss DH, Maasch KA & Anderson DG. (1999). Transitions in the Mid-Holocene. *Science*, **283**, 499-500.

[107] Schnitzler A, Kessler KR, & Benecke R. (1996). Transcallosally mediated inhibition of interneurons within human primary motor cortex. *Experimental Brain Research*, **112**, 381-391.

[107b] Schoenfield HJ. (1965). *The Passover Plot*. The Disinformation Company. NY.

[108] Schuz, A. & Preissl, H., 1996. Basic connectivity of the cerebral cortex and some considerations on the corpus callosum. *Neuroscience and Biobehavioral Reviews* **20**, 567-570.

[108b] Snyder HN. (2000). Sexual Assault of Young Children as Reported to Law Enforcement: Victim, Incident, and Offender Characteristics, National Center for Juvenile Justice, July 2000, *U.S. Department of Justice, Office of Justice Programs*

[109] Semendeferi K & Demasio H. (2000). The brain and its main anatomical subdivisions in living hominoids using magnetic resonance imaging. Journal of Human Evolu tion, **38**, 317-332.

[110] Shiffer F. (1996). Cognitive ability of the right hemisphere:possible contributions to Psychological Function. *Harvard Review of Psychiatry*, **4**,126-138.

[110b] Schiffer, F. (1997). Affect changes observed with right versus left latera visual field stimulation in psychotherapy

patients: Possible physiological,psychological, and thera-
peutic implications. *Comprehensive Psychiatry*, **38**, 289–
295.

[110c] Shaywitz SE, Fletcher JM, Holahan JM, Shneider AE,
Marchione KE, Stuebing KK, Francis DJ, Pugh KR,
Shaywitz BA. (1999). Persistence of dyslexia: the Connecticut Longitudinal Study at adolescence. *Pediatrics*, **6**,
1351-9.

[111] Short RV. (1981). Sexual selection in man and the great
apes. In: *Reproductive Biology of the Great Apes*, C.E.
Graham ed., Academic Press, NY.

[112] Smeyne RJ, et al. (1993). Continuous c-Fos expression
precedes programmed cell death in vivo. *Nature, 363*, 166-
169.

[113] Smith LC & Moscovitch M. (1979). Writing posture, hemispheric control of movement and cerebral dominance in
individuals with inverted and noninverted hand postures
during writing. *Neuropsychologia, 17*, 637-644.

[113b] Snyder HN. (2000).Sexual assault of young children as reported to law enforcement: victim, incident, and offender
characteristics, National Center for Juvenile Justice, July
2000, U.S. Department of Justice, Office of Justice Programs

[114] Sperry RW. (1961). Cerebral organization and behavior.
Science. **133**, 749-1757.

[115] Sperry R. (1968). Hemispheric deconnection and unity in
conscious awareness. *American Psychologist*, **23**, 723-733.

[116] Sperry R. (1982). Some effects of disconnecting the
cerebral hemispheres. *Science, 217*, 1223-26.

[117] Springer SP & Deutch G. (1998). *Left Brain, Right
Brain: Perspectives from Cognitive Neuroscience.* 5th edn.
Freeman, NY .

[118] Stephan KE, Fink GR & Marshall J C. (2006). Mechanisms
of hemispheric specialization: Insights from analysis of
connectivity. *Neuropsychologia, 45*, 209-228.

[119] Stephan KE, Marshall JC, Friston KJ, Rowe JV, Ritzl A, Zilles K, Fink GR.(2003). Lateralized cognitive processes and lateralized task control in the human brain. *Science,* **301**, 384-386.

[120] Stringer C & Gamble C. (1993). *In Search of the Neanderthals.* Thames and Hudson, NY.

[121] Swinburne R. (1986). *The Evolution of the Soul*, Oxford University Press, London.

[122] Swisher CC, Rink WJ, Anton SC, Schawarcz HP, Curtis GH, Suprijo A & Widiasmoro NI. (1996). Latest Homo erectus of Java: Potential contemporaneity with Homo sapiens in Southeast Asia. *Science,* **274**, 1870-1874.

[123] Tannen D. (1986). *That's Not What I Meant.* Ballentine Books, NY.

[124] Tannen D. (1990). *You just don't Understand.* Ballentine Books, NY.

[125] Tannen D. (1994). *Talking from 9 to 5*. Avon Books, N.Y.

[126] Tan U, Tan M, Polat P, Ceylan Y, Suma S & Okur A. (1999). Magnetic Resonance Imaging Brain Size/IQ Relations in Turkish University Students. *Intelligence,* **27**, 83-93.

[127] Terasaki O, & Okazaki M.(2002). Transcallosal conduction time measured by hemifield stimulation of face images. *Neuroreport*, **13**, 97-99.

[128] Torrance EP & Reynolds CR. (1980). Norms and technical manual for "Your style of learning and thinking". : Department of Educational Psychology, University of Georgia. Athens, GA.

[129] Tramo MJ, Loftus WC, Stukel TA, Green RL, Weaver JB, Gazzaniga MS.(1998). Brain size, head size, and intelligence quotient in monozygotic twins. *Neurology*, **50**, 1246-1252.

[130] Trinkhaus E & Shipman P. (1992). *The Neanderthals.* Random House, NY.

[131] Van Kleek MH. (1989). Hemispheric differences in global versus local processing of hierarchical visual stimuli by

normal subjects: New data and a meta-analysis of previous studies. *Neuropsychologia,* **27**, 1165-1178.

[132] Vogt BA, Finch DM & Olson CR. (1992). Functional heterogeneity in cingulate cortex: The anterior executive and posterior evaluative regions. *Cerebral Cortex,* **2**, 435-443.

[133] Wada JA. (1977). Prelanguage and fundamental asymmetry of the infant brain. *Annals of the New York Academy of Science,* **299**, 370-379.

[134a] Wallum H, Lichtenstein P, Neiderhiser JM, Reiss D, Ganiban JM, Spotts EL, Pederson NL, Anckarsater H, Larsson H, & Westberg L. (2012). Variation in the oxytocin receptor gene is associated with pair-bonding and social behavior. *Biological Psychiatry* **71**, 419-426.

[134] Walton ME, Bannerman DM, Alterescu, & K Rushworth MFS. (2003).Functional specialization within medial frontal cortex of the anterior cingulate for evaluation effort-related decsions. *Journal of Neuroscience,* **23**, 6475-6479.

[135] Wasson RG, HoffmanA & Ruck CAP. (1978, 2008). *The Road to Eleusis: Unveiling the Secrets of the Mysteries,* North Atlantic Books, Berkeley, CA.

[136] Weintraub S & Mesulam MM. (1987). Right cerebral dominance in spatial attention. *Archives of Neurology,* **44**, 621-625.

[137] Weisenberg T & McBride KE. (1935). *Aphasia: A Clinical and Psychological Study.* New York: Commonweath fund, (cited in Springer, S.P. and Deutsch, G. *Left Brain, Right Brain: Perspectives from Cognitive Neuroscience.* 5th Ed. p 361, W. H. Freeman, NY, 1999.)

[138] Wegner DM. (2002). The Illusion of Conscious Will. MIT Press, Cambridge, MA,

[139] Westerhausen R, Kreuder F, Dos Santos Sequeira S, Walter C, Woerner W, Wittling RA, Schewiger E, & Wittling W. (2004). Effects of handedness and gender on macro- and microstructure of the corpus callosum and its subregions: a combined high-resolution and diffusion-

tensor MRI study. *Brain Research: Cognitive Brain Research*, 21, 418-426.

[140] Wilkinson DT & Halligan PW. (2003). Stimulus symmetry effects the bisection of figures but not lines: Evidence from event-related fMRI. *Neuroimage, 20,* 1756-1764.

[141] Witelson SF. (1985). The brain connection: the corpus callosum is larger in left handers. *Science* 16, 665-668.

[142] Witelson SF. (1989). Hand and sex differences in the isthmus and genu of the human corpus callosum. A postmortem morphological study. *Brain, 112,* 799-835.

[143] Wittling W. (1990). Psychophysiological correlates of human brain asymmetry: Blood pressure changes during lateralized presentation of an emotionally laden film. *Neuropsychologia, 28,* 457-470.

[144] Wittling W & Pfluger M. (1990). Neuroendocrine hemisphere asymmetries: Salivary cortisol secretion during lateralized viewing of emotion-related and neutral films. *Brain and Cognition, 14,* 243-265.

[145] Wittling W & Roschmann R. (1993). Emotion-related hemispheric asymmetry: Subjective emotional responses to laterally presented films. *Cortex, 29,* 431-448.

[146] Wolford G, Miller MB & Gazzaniga M. (2000), The left hemisphere's role in hypothesis formation. *Journal of Neurosceince, 20*:R64,1-4C

[147] Woodruff PWR, McManus IC & David AS. (1995). Meta analysis of corpus callosum size in schizophrenia. *Journal of Neurology, Neurosurgery, and Psychiatry*, 58, 457-461.

[148] Wrangham RW. (1997). Subtle, Secret Female Chimpanzees. *Science,* 277, 774-775.

[149] Yamaguchi S, Yamagata S, & Kobayashi S. (2000). Cerebral asymmetry of the "top down" allocation of attention to global and local features. *Journal of Neuroscience 20,* RC72, (1-5).

[150] Yazgan MY, Wexler BE, Kinsbourne M, Peterson B & Leckman JF. (1995). Functional significance of individual

variations in callosal area. *Neuropsychologia,* **33**, 769-779.

[151] Yucel M, Stuart GW, Maruff P, Velakoulis D, Crowe SF, Sa vage G, & Pantelis C. (2001). Hemispheric and gender-related differences in the gross morphology of the anterior cingulate/paracingulate cortex in normal volunteers: An MRI morphometric study. *Cerebral Cortex*, **11**, 17-25.

[152] Zeder MA & Hesse B. (2000). The initial domestication of goats (Capra hircus) in the Zagros Mountains 10,000 years ago. *Science,* **287**, 2254-2257.

169 References Total

TWO HUMAN SPECIES EXIST: FAMILIAL POLARITY

GLOSSARY:

Cis-Heterosexuals: A normal sexual identity with their own body sex, and a normal attraction to others of the opposite sex.

Cis-Homosexuals: A normal identity with their own body sex, but attraction to others of the same sex.

Cross-breed: The hybrid, cross-wired offspring from cross-polar couples (see cross-polar)

Cross-polar: Combinations of mates of both familial polarities, Patripolar-Haremic and Matripolar-Orgeic, either as in RM (patripolar)-RF (matripolar) marriage pairs or in LM (matripolar male)-LF (patripolar female) mates.

Dominance in the home: In hemisity complimentary marriages, one parent is the role model, setter and enforcer of standards. This is, by definition, called conditional love. Their children accept, admire, even fear his or her leadership; accept and willingly seek to reach their standards. The other parent is less dominant than the children, and accepts and loves them unconditionally whether they obey or not. The subdominant parent is the child's beloved slave who gives them everything, defends them against the dominant parent's excesses, and can even be abused by the child.

In Haremic families, the patripolar father is the dominant parent. In Orgeic families, the matripolar mother is the dominant parent. In R-R families, an essential subdominant parent is absent, and the children fail to receive truly unconditional love, and may end up angry. In L-L families, the essential dominant parent is absent to set, model, and enforce standards, and the children can become tyrannically dominant over both parents.

Dominance in the workplace and community: Usually the reverse of dominance in the nuclear family. In the workplace, R-bops tend to treat others as equals and to be cooperative, while L-bops must fight for status in a local hierarchy and tend to be combative.

Dyslexia: A learning or cognitive disorder producing challenges with reading, writing, spelling, and second language learning.

Familial Polarity: The two opposite reproductive strategies used by mammals and primates: in Haremic Patripolarity (RM-LF), where males battle for paternity, and in the Orgeic Matripolarity (RF-LM), where sperm battle for paternity. Familial polarity is one of the three personality dualities along with Sexuality, and Hemisity.

Gothic personality: Black adoring, satanic, death oriented, transgender, self-mutilating, LxP

Halo of Life: A circular planar band of area superimosed upon the spiral galactic arms surrounding the black hole of a galaxy within which the potential energy to matter ratio promotes the formation and support of complex beautiful systems, including life.

Haremic: Harem-forming patripolar reproductive strategy, in which males battle for dominance and the females follow the winner-patriarch. Genes of loser males are excluded from the offspring. The patripolar gorillas and orangutans (and Neanderthals) are of the Haremic prototype.

Hemisity:
The two general personality and behavioral styles that result when the brain Executive is inherently either on the right or left side of the brain. Right brain-oriented persons, RPs, have easier access to same side right brain global and emotional skills, and must cross over bridges to access left brain skills, such as language and abstract reasoning. For Left-brain oriented persons (LPs), the reverse is true.

Hybrids of Polarity: The cross-wired offspring of mixed polarity matings, such as L-L or R-R mates. These lack certain properties of purebred polarities, and can be dyslexic or have abnormal sexual orientations.

LF: Left brain-oriented female, Haremic, supportive in family, important details-oriented

Love: The willingness to accept another just the way they are and just the way they are not.

LM: Left brain-oriented male, Orgeic, supportive in family, important details-oriented

LP: Left brain-oriented person, may be male or female, Haremic or Orgeic, supportive in the home

LxP, LxM, LxF: Hybrids from R-R cross polar matings. Cross-wired for sexual identity, thus LxMs are somewhat effeminate. LxF are somewhat masculine.
They are mostly heterosexual,but, have the potential to be bisexual as the sole source of this third sex.

Lumper: An RP who is strong in inductive reasoning and seeing the big picture from important details.

Matripolar: (RF-LM) Right brained dominant females, left brained supportive males, Orgeic (orgy having) reproductive style.
Orgeic: Orgy-having reproductive strategy, in which a matripolar female in heat repeatedly mates with all the males in the troupe. The winner of the resulting sperm race determines the random and unknown father of each child. Prototypes are the Chimpanzees, Bonobos, (and Cro-Magnons). All participating males assume they are the father, resulting in abundant support for the children.

Patripolar: (RM-LF) Right-brain dominant males, Left-brain supportive females, Haremic (harem forming) reproductive style.

RF: Right brain-oriented female, dominant at home, big picture-oriented.

RM: Left brain-oriented male, Haremic, dominant at home, big picture-oriented.

RP: Right brain-oriented person, Orgeic, dominant at home, big picture-oriented.

RxP, RxM, RxF: Right brain oriented hybrid offspring between R-R parents. Developmentally dyslexic but talented. Suffers from mental blocks (temporary decorticate stupidity) when afraid.

Sexuality: One of three dualistic determinants of personality, along with Hemsphericity and Polarity. Sexuality is itself composed of at least three elements: Body sex (M/F), Sexual identity (cis/trans), and Sex of preferred sexual partner (hetero/homo).

Splitter: An Lp with a top down orientation, which dissects wholes into their component parts.

Trans-Heterosexual: Is an Lxp who has the opposite sexual identity than that of their body, but is still heterosexual, although commonly bisexual as well.

Trans-Homosexual: This is a person with the opposite sexual identity of their body sex, who are exclusively attracted to others of their own body sex.

Wild Type: This refers to the original, non-hybrid, non-mutated true-breeding representatives of the two species. They are the patripolar RM and LF, and the matripolar RF and LM.

TWO HUMAN SPECIES EXIST: FAMILIAL POLARITY

APPENDICES

To test others, scan and copy these questionnaires:

Appendix A:　　THE POLARITY QUESTIONNAIRE

Professor Bruce E. Morton: University of Hawaii School of Medicine
Published in Brain and Cognition 49, 53-72 (2002).

Name or I.D.#_____. Sex_, Age__, Handedness _,
Ethnicity of Mother's Parents_____&_____,
Ethnicity of your Father's family_____&_____.

___ 1. When I become upset, after cooling down I don't need to talk, I need to be alone.
___ 2. I tend to be introspective, self-conscious, thin-skinned, and psychological.
___ 3. I would rather maintain and use good old solutions than find new, better ones.

___ 4. I talk more about thoughts, things, or acquaintances than entertainment, sports, or politics.
___ 5. I am comfortable and productive in the presence of disorder and disorganization.
___ 6. I find it very difficult to tolerate when my mate (or important other) becomes defiant to me in private.

___ 7. I don't need a lot of physical contact from my mate.
___ 8. I like daily small reassurances of my mate's love more than monthly large rewards.
___ 9. I tend not to be very romantic or sentimental.

___10. I am more strict than lenient with our children (or I would be if I had children).
___11. Given the opportunity, I am more of an early morning person than a late night person.

Polarity Questionnaire Key:
1. The questions alternate between R and L brain orientations when marked True.
   a. Thus, when odd numbered questions are marked True and even numbered marked False, the result is 11 possible LEFT brain-oriented answers.
   b. Conversely, when even numbered questions are marked T and odd numbered marked F, the result is 11 possible RIGHT brain-oriented answers.
   c. *The questionnaire is scored as the number of Left brain-oriented answers out of 11:* Thus, 1-4 Left answers/11 = Right hemisity, 5 -11/11 Left answers = Left hemisity.

2. Hemisphericity is genetic. You are "either Right, or Left", not a gradation in between.

235

APPENDIXES

Appendix B: # THE ASYMMETRY QUESTIONNAIRE

Professor Bruce E. Morton: University of Hawaii School of Medicine
Published in Brain and Cognition 51, 372-374 (2003).

Your Name or Number:_____Sex___Age___Grandparent's Ethnicity_____

For each of these 15 pairs of statements, mark an X at the START of the ONE statement is MOST like you.

	_____Statement A	Statement B
1.	I often talk about my and other's feelings of emotion.	I tend to avoid talking about emotional feelings.
2.	I am good at finishing projects.	I am a strong starter of projects.
3.	I organize parts into the whole (synthetic, creative).	I break the whole into parts (reductive-reductionistic).
4.	I am quick-acting in emergency.	I methodically solve problems by process of elimination.
5.	I think and listen interactively-vocally, and talk a lot.	I think and listen quietly, keep my talk to a minimum.
6.	I don't read other people's mind very well.	I am very good at knowing what others are thinking.
7.	I see the big picture (project data beyond, can predict).	I am analytical (stay within the limits of the data).
8.	I tend to be independent, hidden, private, & indirect.	I tend to be interdependent, open, public, & direct.
9.	I usually design original outfits of clothing.	I dress for success and wear high status clothing.
10.	I need to be alone and quiet when upset.	I need closeness and to talk things out when upset.
11.	I praise others, and also work for praise from others	I do not praise others, nor need the praise of others.
12.	I'm more interested in objects and things.	I tend to be more interested in people and feelings.
13.	I seek frank feedback from others.	I avoid seeking evaluation by others.
14.	I often feel my mate talks too much.	I feel my mate doesn't talk or listen to me enough.
15.	I'm strict, my kids obey me and work for my approval.	I'm not a strict parent, my kids don't obey me well.

L Score = Even As 7 + Odd Bs 8 = 15 Ls / 15 Ls. Right Hemisity =5 or less L answers. Left Hemisity = 6 or more L answers

236

Appendix C: **Hemisity Questionnaire**

Bruce E. Morton, Ph.D., University of Hawaii School of Medicine

Name or I.D.#_____. Sex___, Age___, Handedness___, Ethnicity of your Mother's family_____, Ethnicity of your Father's family_____.

Write an A or B for the statement most like you, or most like the way you think.

___1. After I have been upset with my mate, **A**. I need to be alone and quiet, vs. **B**. I need closeness and to talk things out. (If you are not currently in such a relationship, imagine how you would feel if you were.)

___2. If my mate defies me in private, I find it to be, **A**. very difficult to tolerate, **B**. something I can put up with.

___3. **A**. I am analytical (stay within the limits of the data), vs. **B**. I see the big picture (predict beyond data).
___4. Regarding disorder, **A**. I am stressed and slowed by it, vs. **B**. I am comfortable or accelerated by it.

___5. **A**. I often feel my mate talks too much, vs. **B**. I feel my mate doesn't talk or listen to me enough.
___6. **A**. I often talk about my and other's feelings of emotion, vs. **B**. I tend to avoid talking about my or other's emotional feelings.

___7. **A**. I tend to be independent, hidden, private, indirect, vs. **B**. I tend to be interdependent, open, public, and direct.
___8. In this country I wish there were, **A**. more high-quality law and order, or **B**. more personal freedom.

___9. **A**. My daydreams are not vivid, vs. **B**. My daydreams are vivid.
___10. **A**. My thinking consists of images, vs. **B**. My thinking often consists of words.

___11. I feel that I am more, **A**. conservative and cautious, vs. **B**. innovative and bold.

___12. As a parent in a nuclear family, I am, **A**. the most dominant spouse, vs. **B**. the most supportive spouse (If not presently in a spousal relationship, imagine that you were in one.)

___13. Can you comfortably carry on a conversation with someone in the same room and with another person on the telephone at the same time? **A**. No, vs. **B**. Yes.

___14. To others my desk might appear to be, **A**. neat, vs. **B**. messy.

___15. When relating to others I would describe myself as, **A**. sensitive, vs. **B**. intense.

___16. In terms of my health, **A**. I am almost never ill, vs. **B**. I catch colds, the flu, etc., rather easily.

___17. If I were to self-medicate with drugs, I would choose, **A**. a depressant such as alcohol or cannabis vs. **B**. a stimulant such as cocaine or amphetamine.

___18. **A**. I often enjoy chatting with others, vs. **B**. I tend to find social chatter to be somewhat annoying.

___19. **A**. I don't read other people's minds very well, vs. **B**. I am good at knowing what others are thinking.

___20. **A**. I tend to take the blame, vs. **B**. I try to avoid taking the blame.

___21. **A**. I avoid deeply experiencing or expressing my emotions because they seem so overwhelming I am afraid I might lose control.   **B**. I am not afraid to deeply experience and express my emotions because they are not that overwhelming.

Odd As + Even Bs = Left score 21/21. Right hemisity = 8/21 Left answers or less. Left hemisity = 9/21 or more Left answers.

Appendix D.   A BINARY PREFERENCE QUESTIONNAIRE
Bruce E. Morton, Ph.D., University of Hawaii School of Medicine

Your Name or Number:_____, Sex, M or F___, Handedness, R or L__, Parental Ethnici-ty_____,_____.
For each pair of statements, mark an X by the viewpoint that is most like your own.

Statement A	MEMORY PROCESSING	**Statement B**

1. I look for differences, separate, and analyze things.   I look for similarities, commonalities, and unify things.
2. I organize parts into the whole (synthetic, creative).   I break the whole into parts (reductive, reductionistic).
3. I manipulate concepts deductively, see important details. I manipulate contexts inductively, can generalize.
4. I see the big picture (project data beyond, can pred   I am analytical (stay within the limits of the data).
5. I symbolize and label things: (a symbol=1000 words).   I visualize things: (a picture=1000 words).
6. I imagine, convert concepts into contexts or metaphors.   I use logic, convert objects into literal concepts.

TYPE OF CONSCIOUSNESS
7. I thrive in the early morning.   I am alert in the late evening.
8. I am good at completing things.   I am a strong starter of projects.
9. I can easily concentrate on many things at once.   I tend to concentrate on one thing in depth at a time.
10. I am orderly, organized, and deliberate.   I am disorganized, disorderly, but fast.
11. I am quick-acting in emergency.   I methodically solve problems (process of elimination)
12. I am uncomfortable with chaos, and am slowed by it.   I am comfortable with chaos, am accelerated by it.
13. I have good ideas, not all of which are practical.   I'm very intuitive, insightful about idea applications.
14. I'm self-conscious, feel guilty, and am a poor performer. I' not self-conscious, have low guilt, & perform well.
15. I don=t read other people=s mind very well.   I am very good at knowing what others are thinking.
16. I feel communication is my source of power and support. I feel communication is less important to me.

FEAR, AROUSAL, SENSITIVITY
17. I value tradition, respect authority, and resist change. I am innovative, question authority, and seek change.
18. I am more radical, daring, and experimental.   I am more conservative, cautious, and avoiding.
19. I tend not to invade other's boundaries.   I may invade other's boundaries.
20. I often talk about my and other feelings of emotion.   I tend to avoid talking about emotional feelings.
21. I tend to be independent, hidden, private, and indirect.   I can be interdependent, open, public, and direct.
22. I seek frank feedback from others.   I avoid seeking evaluation by others.
23. I am comfortable in groups, even adversarial ones.   I am uncomfortable in groups, unless loyal friends.
24. I have an out-of-control temper, but it only lasts minutes.   I can control my anger but it may last for hours.
25. I need to be alone and quiet when I am upset.   I need closeness, to talk things out when I'm upset.
26. I often take responsibility, blame myself, or apologize.   I usually avoid taking the blame.
27. I'd rather rationalize a way to be right than be wrong.   I'd rather be wrong than rationalize a way to be right.

GENERAL BEHAVIORAL STYLE
28. I think and listen interactively-vocally, and talk a lot.   I think and listen quietly, keep my talk to a minimum.
29. I tend to use humor to tease or mock the other person.   I often tend to make humor at my own expense.
30. I praise others, and also work for praise from others.   I do not praise others, nor need the praise of others.
31. I'm immediate, thick-skinned, no time for self analysis. I'm contemplative, thin skinned, intense self-analysis.
32. I usually design my own outfits of clothing.   I dress for success and wear high status clothing.
33. I'm more interested in objects and things.   I tend to be more interested in people and feelings.
34. I'm very observant and in touch with my surroundings.   I'm often thinking & tend to ignore my surroundings.
35. I often feel my mate talks too much.   I feel my mate doesn't talk or listen to me enough.
36. I am the nurturance-requiring member of a couple.   I am the more nurturing member of a couple.
37. I am a supportive, highly competitive partner.   I am an innovative, directive, yet cooperative partner.
38. I'm strict, my kids obey me, and work for my approval.I'm not a strict parent and my kids don't obey me well.
39. I don't need a lot of physical closeness from my mate.   I need lots of physical closeness from my mate.
40. I find it intolerable if my mate defies me in private.   I can tolerate it if my mate defies me in private.

Appendix D. continued:

%Left Score = Odd Statement As <u>20</u> + Even Statement Bs <u>20</u>

= 40 Left answers/40

Right Hemisity = 14/40 or less Left answers.

Left Hemisity = 15/40 or more Left answers.

APPENDIXES

INDEX

TWO HUMAN SPECIES: FAMILIAL POLARITY

ABOUT THE AUTHOR:

Bruce Eldine Morton was born Southern California in 1938. After Completing the M.S. and Ph.D. degrees in biochemistry at the University of Wisconsin, he spent post-doctoral periods as a Research Fellow at Wisconsin's Institute for Enzyme Research, M.I.T., and Harvard Medical School. He was hired by the School of Medicine at the University of Hawaii, 1969 where he directed a neuroscience research laboratory long after his "retirement" in 1995.

In 1974, Dr. Morton set a world distance record in a hang glider. He has also been active in gymnastics, SCUBA diving, wind surfing, snowboarding, and now with dual purpose motorcycle riding to Mayan ruins in Guatemala. He has been a member of many choral societies and performed in concerts with the Boston Symphony.

Dr. Morton spent sabbaticals at USC, Stanford, and at the University of Michigan. He is also a member numerous professional societies, including the International Society for Research on Aggression. His most recent publication was #80: BE Morton and SE Rafto, Behavioral laterality Advance: Neuroanatomical Evidence for the Existence of Hemisity, *Personality and Individual Differences 49,* 34–42 (2010). From his home base in Guatemala, Dr Morton continues research upon the removal of psychological stress. He may be contacted at bemorton@hawaii.edu or www2.hawaii.edu/~bemorton.

ABOUT THE AUTHOR

www.ingramcontent.com/pod-product-compliance
Lightning Source LLC
Chambersburg PA
CBHW021616270326
41931CB00008B/726